A STRANGE ENGINE OF WAR

A STRANGE ENGINE OF WAR

*The "Winans" Steam Gun and
the Civil War in Maryland*

JOHN W. LAMB

Chesapeake
BOOK COMPANY

BALTIMORE, MARYLAND

2011

Copyright © 2011, John W. Lamb
Published by The Chesapeake Book Company
112 Elmhurst Road, Baltimore, Maryland, 21210

Library of Congress Cataloging-in-Publication Data

Lamb, John, 1968-
A strange engine of war : the Winans steam gun and the Civil War in
Maryland / by John Lamb.
    p. cm.
Includes bibliographical references and index.
ISBN 978-0-9823049-2-1 (alk. paper)
1. Steam guns (Ordnance)--History. 2. Winans, Ross, 1796-1877. 3.
Dickinson, Charles S. 4. Maryland--History--Civil War, 1861-1865--
Artillery operations. I. Title. II. Title: Winans steam gun and the Civil War
in Maryland.
UF630.L23 2011
623.4'2--dc23

                        2011017872

Manufactured in the United States of America.
The paper used in this publication meets the minimum requirements
of the American National Standard for Information Sciences
Permanence of Paper for Printed Library Materials
ANSI Z39.48-1984.
Printed on recycled paper

Designed by James F. Brisson

⊰ CONTENTS ⊱

⇥ LIST OF ILLUSTRATIONS ⇤

⊰ACKNOWLEDGMENTS⊱

My interest in Ross Winans began in the early 1990s. While working on another project, I became intrigued by the gun and wondered how it worked. At first the facts seemed clear, but questions began to arise. Did Winans really invent the gun, did it work, and what became of it? Finding the answers took much longer than I ever imagined. Along the way several people made important contributions: Dr. Benjamin McArthur, then of Southern Adventist University, helped lay the foundation for this work in his research methods class by recommending shelf-reading as a useful research tactic. (When a search in the most logical source fails, look to books on the shelves near it.) This tactic, applied to books and then to online research, led to the discovery of crucial information in unexpected places. My wife Cindee put up with my obsession about the gun over the years as the work was being completed. Dan Hunter and Ralph Osgood located and graciously shared important period documents surrounding the gun's capture and where it was eventually taken. Historian Charley Mitchell read several drafts and helped bring focus to a complex tale.

This is the story of a weapon that disappeared long ago, but it speaks to deeper issues of truth and memory. I hope you enjoy reading it as much as I have delighted in uncovering it.

John W. Lamb
Chattanooga, Tennessee, 2011

❧ INTRODUCTION ❧

The Birth of a Legend

In April 1861, with South Carolina and six states of the Deep South having seceded from the Union and four others considering secession, President Abraham Lincoln called for 75,000 volunteers to put down the rebellion and, most pressingly, to secure the nation's capital. On April 19, 1861, as a contingent of Massachusetts volunteer militia passed through the historically tumultuous city of Baltimore, they were attacked by a pro-secession mob. Troops and rioters exchanged fire, and both sides suffered casualties. In the wake of what quickly became known in the North as the Baltimore Riots and in the South as "the Lexington of 1861," newspapers across the country also described a strange and mysterious weapon authorities had discovered in Baltimore.

Over the years, the story of that weapon found its way into countless books and articles. Ross Winans, a wealthy builder of railroad locomotives and allegedly a "rabid secessionist," was said to have created a steam-powered mechanical gun to blast the ranks of Union troops should they seek to subdue Baltimore. Rumors spoke of its fearsome rate of fire and its ability to move on its own. After putting the gun on public display at the height of the panic following the riot, Winans supposedly tried to smuggle it out of Baltimore and into the arms of Confederate forces gathering at Harpers Ferry. Instead, Federal troops descended on what is now Ellicott City and captured the machine, only to find it to be unworkable because key parts were missing. The gun was never tested, and its true capabilities remain a mystery to this day.

It is a colorful tale to be sure, but period press accounts and documents reveal a far different reality. Myths and misrepresentations have obscured the gun's true origins in Ohio in the 1850s, how it came to be in Baltimore in April 1861, what later became of it, and, most important, how little any of this had to do with Ross Winans.

Steam-powered devices similar to the steam gun were familiar parts of life in the 1850s and 1860s. Here steam-powered fire engines join in the celebration at the opening of the Ohio and Mississippi Railroad in Cincinnati in 1857. (Frank Leslie's Illustrated Newspaper, *June 6, 1857 / Library of Congress.*)

❧ 1 ❧

A Flowering of Invention

STEAM POWER was the driving force behind the industrial revolution, and American inventors spent prodigious energy attempting to harness it for military purposes. The first and perhaps best known of these devices, aside from the "Winans" Steam Gun, was that invented by Jacob Perkins of Massachusetts, who worked as an engraver of coins and who developed a process for using steel plates to print bank notes. Unable to find financial support for his inventions, he moved to London, where he created others, including a high-pressure steam engine. He then turned his attention to a steam gun, which he patented in Britain in 1824.

His device required a high-pressure boiler for steam. The gun was fed by a hopper and could fire shot through eleven one-inch planks, as well as steel plate a quarter of an inch thick, from a distance of thirty-five yards. It is reported to have fired as many as one thousand rounds per minute. British and French military officers examined the gun, but neither nation adopted it for its army. Despite its rapid-fire capability, its use in the field was limited by the need for a high-pressure boiler of the kind generally found only on ships or in factories. Although the gun was never adopted by the European military, it was a star attraction at the National Gallery of Practical Science that Perkins opened in London, where it was fired down the length of the gallery's main room.[1]

The Perkins Gun and other attempts at a "steam gun" sought ways to use the expansion of steam to throw projectiles, but an alternative approach soon appeared. Perhaps looking to the sling shot, inventors envisioned hand-cranked or steam-powered devices that would supposedly harness what they styled centrifugal force for destructive purposes.

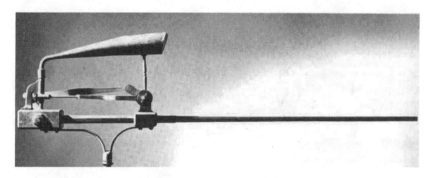

Jacob Perkins obtained a patent for this steam gun in Great Britain in 1824. It was said to be capable of firing shot through eleven one-inch planks and a steel plate at a distance of thirty-five yards. (Photograph Courtesy of Baker Perkins Historical Society.)

Though the use of the term centrifugal is of the period, more current physics terminology would describe the force involved as centripetal. It is centripetal force that acts on the projectiles in an old-fashioned sling shot. When swung in a circle, the shot is pulled towards the center by centripetal force in the form of the sling. When the sling shot is released, the centripetal force ceases and the projectile travels away in a straight line. Were a person in the sling shot, he would feel a powerful force pulling him outward, away from the center. This is "centrifugal force," a virtual force caused by an object's inertia when diverted by centripetal force from its normal straight line of travel.[2]

Thus, while inventors of these weapons thought they were harnessing the power of "centrifugal" force, the success of their weapons depended not on this "virtual" force but on how much centripetal force their devices could transmit to shot before it was released in the direction of the target. They failed to grasp the fact that no amount of gearing could spin the barrel or disk of a "centrifugal gun" fast enough to launch a projectile as fast, as far, or with as much killing power as that created by the combustion of gunpowder in the barrel of a pistol, musket, or cannon.

Basic principles of physics aside, even if these devices had potential, America's military establishment was very much aligned with European tactics in the use of infantry and artillery. Even promising improvements in conventional weapons were not readily considered. This reluctance on the part of army officials made it difficult for inventors of unconventional devices even to have the opportunity to make a case for their machines, and it helps to explain the dramatic demonstrations they arranged to gain notice for their work.

What follows is a brief look at some other "centrifugal" guns invented between 1838 and 1861. The sketches are drawn mostly from patent records and include several guns mentioned in period scientific publications.

Robert McCarty's Centrifugal Gun

AUGUST 1838 found Robert McCarty of New York filing for a patent on a "Machine for Throwing Balls, Shot, Etc." His would become the first patent in the United States for a centrifugal gun.

McCarty's "machine" featured a round, vertically mounted housing that was in two pieces. "Scrolls" inside each piece formed a "barrel" that ran in an arc from the center to the "muzzle" of the weapon. As projectiles were fed into the gun, its "propelling arm" pushed shells into the groove between the scrolls and then pushed them out as it rotated. McCarty's patent application noted that the gun could be operated in a horizontal position but that he preferred to use it vertically, as shown in the drawings in the patent.[3]

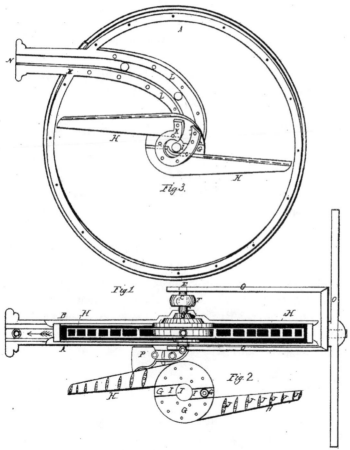

The gun designed by Robert McCarty, patented in 1838 and tested in New York in 1861. It was the first to receive a U.S. patent and was said to have fired 480 rounds per minute "in a perfect stream" with a range of nearly a mile. (United States Patent, No. 1,049.)

May 15, 1861, found it being tested "at the foot of 39th Street, North River" in New York. "It is one of the most singular implements of war that has ever been exhibited to the American people," the *New York Herald* reported, adding that it placed the "Winans' gun "entirely in the shade, sending balls at the rate of 480 per minute without any powder or apparent effort." Manned by a team of six men "at the cranks," a seventh feeding shot to the gun, and the inventor, "balls poured out of the gun in a perfect stream, and it appeared that one continuous stream was being hurled against the target." That target consisted of three wooden boards about fifty yards away. The gun was then aimed across the North River. Balls dropped into the water just short of the far bank, about a mile away.

During the test the gun fired one-inch shot, but McCarty was also working on a steam-powered gun to throw thirty-two-pound shot. The smaller gun had been built by J. Cowell of No. 340 West Twenty-Fourth Street, who was ready to turn out several a week. The account noted that an attempt was being made to interest the War Department in it, and that an army officer who had witnessed the test was thinking of securing a gun for his regiment.[4]

Benjamin Reynolds's Centrifugal Gun

BENJAMIN REYNOLDS of Kinderhook, New York, also invented a centrifugal gun in the late 1830s. His device was operated by two men, one on each side operating a crank. A hopper dropped shot into a revolving drum that threw them. Reynolds demonstrated his gun at West Point in 1837, where it sent a thousand two-ounce shot through 3.75 inches of solid pine at a distance of 110 yards. Reynolds later took it to Washington, where further demonstrations were made before a congressional committee and military officers. The Washington tests used a target of three one-inch pieces of pine at 150 yards. The shot from the gun went through the target and fell into the Potomac River 300–400 yards beyond. It fired so quickly that the committee could not determine what *fraction of a second* was required to fire sixty balls.[5]

Joseph Martin's Central Power Engine (1840)

DR. JOSEPH MARTIN of Louisville, Kentucky, received a patent for what he called a "Central Power Engine" that would employ the "centrifugal principle" to throw projectiles or drive machinery. Martin's "engine" used two sets of rotating chains with buckets joined by gearing to throw projectiles. Interestingly, he suggested that his gun could be water-powered.[6]

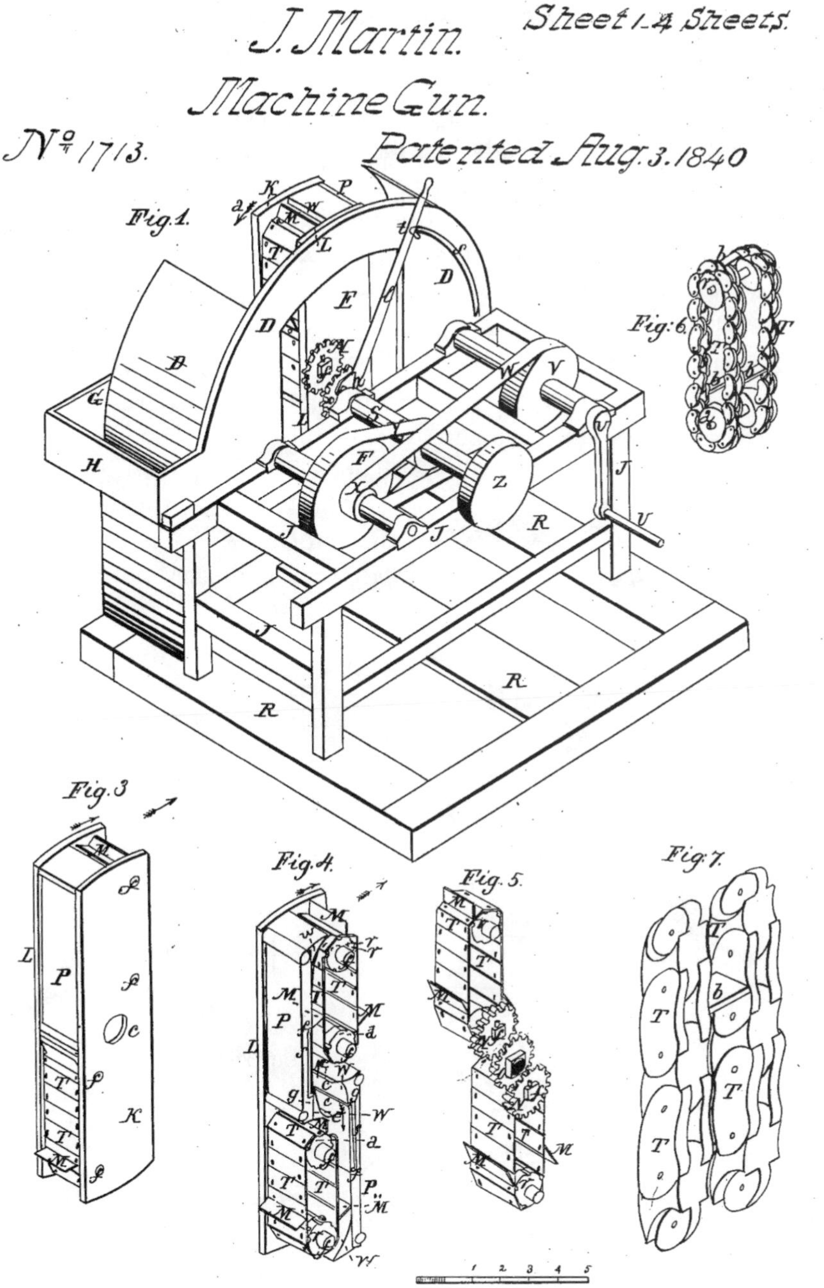

J. Martin.

Machine Gun.

Nº 1713.

Sheet 1.4 Sheets.

Patented Aug. 3. 1840

Fig. 1.

Fig. 6.

Fig. 3

Fig. 4.

Fig. 5.

Fig. 7.

Dr. Joseph Martin's "Central Power Engine," designed in 1840, could be used to fire projectiles or drive machinery. (United States Patent, No. 1713.)

Smith & Weaver's Improved Machine for Throwing Projectiles

A. B. SMITH AND WILLIAM WEAVER of Clinton, Pennsylvania, received a patent for their "Improved Machine for Throwing Projectiles" on August 19, 1856. Their device featured a horizontally mounted, ring-shaped shield. Inside was a rotating disk. Shot was dropped into a hopper and directed by a tube into the rotating disk. The gun employed a "valve" that allowed the operator to control the rate of fire from one shot per revolution of the disk to a continuous stream.[7]

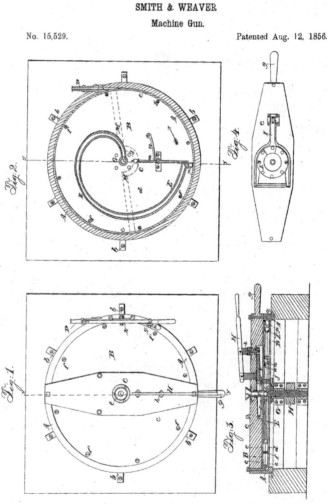

SMITH & WEAVER

Machine Gun.

No. 15,529.   Patented Aug. 12, 1856.

A. B. Smith and William Weaver of Pennsylvania received a patent for this device in 1856. (United States Patent, No. 15,529.)

Albert Potts's Centrifugal Battery

ALBERT POTTS of Philadelphia received a patent for his "Centrifugal Battery" on May 19, 1857. His design began with an upright wheel mounted on an axle held by two A-shaped supports. Force was applied by a pulley on the right side of the axle. Projectiles were fed through the center of the axle on the left side of the device, and passed into a channel inside the wheel. A timing device activated by the rotation of the wheel helped ensure that projectiles were thrown one by one at the proper point in the wheel's rotation. The angle at which projectiles left the device was adjustable.[8]

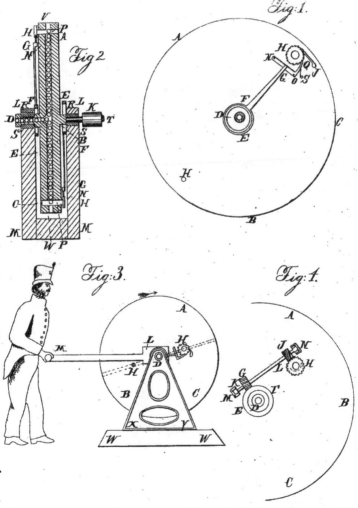

Albert Potts's "Centrifugal Battery," 1857. (United States Patent, No. 17,339.)

C. B. Thayer's Centrifugal Gun

C. B. Thayer of Boston's idea of an "Improvement in Centrifugal Guns" was patented in August 1858. It featured a large ring atop a solid base. Inside this ring, a barrel assembly shaped loosely like an "S" rotated horizontally. When shot was dropped into a hopper, it fell into the center of the barrel. The tip of the rotating assembly was in close proximity to the outer ring, and as it rotated, the shot was thrown out through an opening in the ring. The gun was designed to be run horizontally, or vertically if a target was at long range.[9]

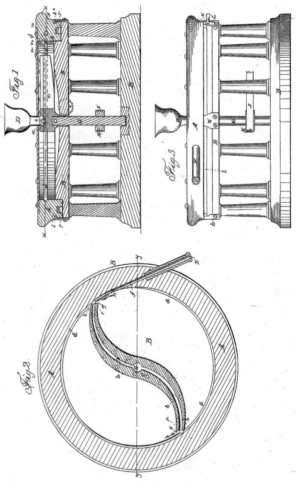

C. B. Thayer's "Improvement in Centrifugal Guns," 1858, was similar to Smith and Weaver's in design but instead of a paddle-like device to push projectiles, it had an S-shaped assembly that pushed projectiles along the inside of a circular shield. (United States Patent, No. 21,109.)

William Joslin's Centrifugal Gun

WILLIAM JOSLIN's "Improvement in Centrifugal Guns," patented in May 1859, was designed to be cranked by hand. Two gears set it in motion—one connected to the barrel assembly and one on which the assembly rotated. A hopper fed balls into the assembly. As the assembly rotated, a wedge triggered a slide, allowing a ball to escape. The faster the crank was turned, the more balls the gun would throw.[10] Although the Joslin gun has been overlooked by historians, its story is intertwined with the device that would come to be known as the "Winans" Steam gun and will be fully explored later.

Sir McDonald Stephenson's Steam Mortar

IN JUNE 1860, a description of yet another centrifugal gun appeared in newspapers in Wisconsin, California, and probably elsewhere in the United States. In Britain, Sir McDonald Stephenson was at work on a gun driven by steam. He proposed to "apply a rotary steam engine to impart high centrifugal force to a cylinder about seven feet in diameter, near the center of which the shot are inserted, and pass down the arms or spokes to the periphery, where they are retained by the apparatus, which is regulated to release them at the precise time required." According to Stephenson, his gun would be able throw the heaviest of shot, and the power, velocity, range, rate of fire and angle of fire could be adjusted as necessary.[11]

Dr. Draper Stone's Steam Gun

In May 1861, a gun designed by Dr. Draper Stone drew notice. It consisted of a boiler and engine with a revolving barrel, the length of which varied depending on the size of projectile desired. The barrel had six openings from which shells would be forced out as fast as it could revolve—between two and three hundred times per minute "without any difficulty." It could be aimed in any direction and could be made to throw musket balls two miles "and kill a man." The inventor claimed it could also be constructed at a size large enough to send a five-pound shot five miles.[12]

These and perhaps other as yet unidentified pre-war approaches to "centrifugal" guns, are a testament to their inventors' ingenuity and form the backdrop for the story of what arguably would become the most famous weapon of its kind.

Machine Gun.

No. 24,031. Patented May 17, 1859.

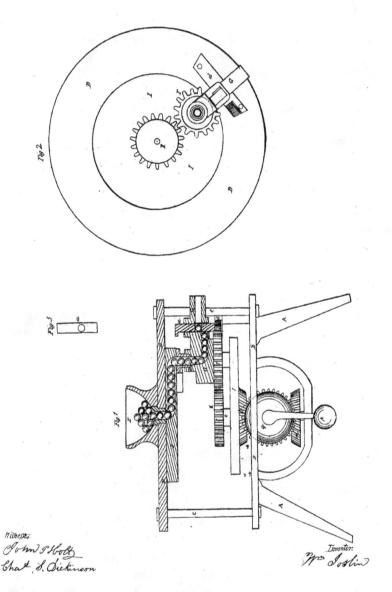

Original schematic of William Joslin's "Centrifugal Gun," developed in Waterford, New York, and Cleveland. Joslin's gun resembles a coffee grinder and was developed through a partnership between Joslin and Charles S. Dickinson. Angered by his omission from patent documents, Dickinson set out to patent his own gun, which became known as the Winans Steam Gun. (United States Patent, No. 24,031.)

⊰ 2 ⊱

William Joslin and
Charles S. Dickinson
Build a Centrifugal Gun

Among many steam and centrifugal guns, the device invented by William Joslin marks the beginning of the history of the gun that gained prominence in Baltimore in 1861. Piecing together the story of Joslin's gun requires reading between the lines of a series of angry letters to the editor of the *Cleveland Leader* from Joslin and others associated with the gun's creation, who were involved in a public dispute over its true inventor.

Joslin, a mechanic, had been at work since 1851, if not before, on a hand-powered centrifugal gun, in Waterford, New York, and later in Cleveland. He kept a small model of the gun in his Cleveland shop from 1856 onward.[13]

According to Joslin, Charles S. Dickinson, a dancing instructor and gymnasium operator, brought a "Mr. Weaver" to his shop in November or December 1858 to discuss developing it. "They were to furnish money to try it on a larger scale than the model I had in my shop at the time." Joslin entered into an "arrangement" with them, and they built a "machine" in his shop. "Upon the experiments of this machine Mr. Weaver backed out, and let Mr. Dickinson have his interest. Instead of these parties going to the expense of this experiment, as was agreed on in my shop, I paid one-half the expense myself," he wrote.[14]

Dickinson and Joslin then built a second prototype in Joslin's shop. "This machine was taken into Dickinson's gymnasium in the spring of 1859. At this point," Joslin maintained, "Dickinson began claiming that the invention was his, and had an article published counting himself as the inventor. Here commenced the piratical course of Mr. Dickinson, in relation to me."

According to Joslin, Dickinson proposed having the patent papers come out without Weaver's name, since Weaver no longer had an interest in the gun. A Mr. Rider was engaged to copy the required papers but altered them to cover any machines that might be invented for use in building the guns. Joslin refused to sign them. Although those papers were drafted, the patent for the gun was issued in Joslin's name only.[15]

Joslin's "Improvement in Centrifugal Guns," patented in May 1859, was a hand-cranked device. Two gears set it in motion, one connected to the barrel assembly and one on which the assembly rotated. A hopper fed balls into the assembly. As the assembly spun, a wedge triggered a slide, allowing a ball to escape. The faster the crank was turned, the more balls would be thrown.[16]

Dickinson was apparently angered by the fact that the patent did not bear his name. "Upon this Mr. Dickinson commenced defaming me in an unjust manner, and declared I never should make anything out of my invention," Joslin complained in 1861. "He has for nearly two years carried on an unjust course against me, and the only knowledge I have had of it is what I derived from the newspapers."[17]

Exhibitions of the model gun were undertaken. "And here [Cleveland], as elsewhere, and at all places, wherever exhibited, Mr. Dickinson endeavored to sell so much of the right to the patent as to enable him to build a gun," wrote L. J. Rider, the attorney responsible for the contracts between the gun's partners. ". . . After a time Mr. Barker was induced to and did furnish money wherewith Mr. Dickinson went to Boston, he having therefore patented an improvement on said invention, with the understanding that the whole thing should be offered to the Government, then to France or any foreign Government."[18]

Dickinson's "improvement," patented in August 1859, replaced Joslin's complicated hand-driven gearing with a balanced barrel, much like an extra-large musket barrel, that would be spun by a steam engine. As Dickinson described it: "In operating this gun the barrel is set in motion (a circular motion being given to it) by means of steam or other power, said barrel being connected with other machinery, of course, such as gear-wheels, &c., which are not here represented. The balls pass down through the bore in the upper part of the shaft . . . connecting with the bore of the gun, and then into the gun-bore."

A hinged lever was anchored on one side of the bore opening. Fitting around the opening of the barrel, the lever was attached to a spring that held it down, blocking the barrel with a pin. When the barrel was spinning fast enough, force would be applied to a thin rod running through the center of the gun's axle. This rod was connected to the underside of the hinged lever. It would push the lever up and unblock the barrel for a

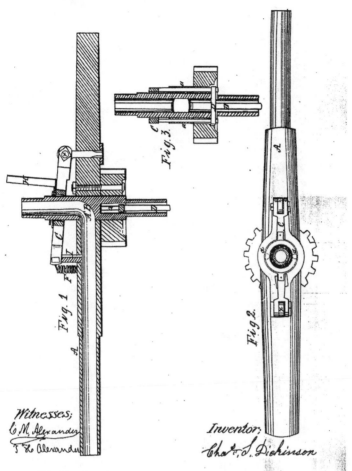

The gun as modified by Charles S. Dickinson. Though often lamented as lost, the details of the most crucial part of the "Winans" Steam Gun—its barrel assembly—are to be found in Charles Dickinson's patent from 1858. (United States Patent, No. 24,997.)

shot to be thrown out. "The operation of the rod . . . is only momentary, and consequently, the lever . . . and the pin pass back to their positions almost instantly, allowing one ball to pass out through the barrel," wrote Dickinson. That particular device was the heart of Dickinson's invention. As the gun's remaining parts—the boiler and gearing that would drive the barrel—were not unique, they were not detailed.[19]

The *Milwaukee Sentinel* of March 5, 1860, contained a description of what probably was one of the prototypes Dickinson and Joslin built, which at the time was on display at the billiard table factory of "Dickinson & Company." The device was likened to an "immense coffee roaster with a crank on one side, and a narrow opening on the opposite." Inside, "a couple of barrels, made to revolve by the outside crank, [were] fed with bullets from a hopper on the top." The writer was present when the gun was tested and reported that it "perforated" a wooden plank an inch thick. The person demonstrating the gun claimed that when steam power was applied to the device it would "send bullet after bullet with so nice an accuracy that they would all disappear in one spot," a claim the reporter regarded with skepticism. In June 1860, the *Sentinel* referred to this demonstration in connection with a mention of Sir McDonald Stephenson's steam mortar, adding that the earlier demonstration had been conducted by "Col. Dickinson of the *Albany*." A later clipping noted that "Col. Dickinson" was the brother of the Dickinson who invented the steam gun connected to Baltimore.[20]

In August 1860, a centrifugal gun described as an "infernal machine" was exhibited in Columbus, Ohio. It was said to be capable of throwing five hundred balls per minute by the turn of a crank, and the demonstration was noted by papers in Ohio, Georgia, and California.[21] Though not expressly linked to Dickinson, it might also have been a demonstration of his gun.

Sometime in 1860, Dickinson built a full-scale working version of his steam-powered gun "at his own expense" in Boston. In November of that year he took it to Baltimore, he would later explain, "in order to be near the national capital, hoping thereby to have the opportunity of exhibiting it and of having its merits tested by a competent committee of officers of the United States Army."[22]

Though Charles Dickinson said the gun was in Baltimore after November 1860, papers in Ohio, Louisiana, and Wisconsin noted his demonstration of a hand-cranked centrifugal gun in Boston in late December. It was probably the earlier model constructed in Cleveland. The gun was said to be able to "throw five hundred balls per minute at a long rifle range, without powder or cap, simply by turning a crank like a coffee mill."[23]

Meanwhile, in Baltimore, according to Dickinson, the steam gun was "put into successful operation before large numbers of citizens and several officers. Many notices of it were sent to the War Department upon which no official action was ever taken," he wrote in August 1862. "It was proved at the exhibition that it would throw to a distance of one mile with precision and effect from four hundred to five hundred two ounce balls per minute, with a horizontal sweep of one hundred and twenty degrees."[24] Undoubtedly, the exhibition was designed to attract investors, and there are indications that a circular advertising it was being distributed in the city.[25]

Were it not for a quirk of city law, the location of these demonstrations might be unknown, but permission was required from the city council for the use of a steam engine within the city. At the council's January 28, 1861, session, the *Baltimore Sun* reported, "Mr. Brown presented the petition of David Dickinson to have a steam-engine on his premises, No. 24 North Street for the purpose of exhibiting a patent centrifugal gun." The matter was referred to the fire department committee. Notice that "David Dickerson" had received the council's approval appeared in the paper on February 5, 1861. In another portion of the same paper, "David Dickinson" was granted permission to temporarily use a four-horsepower steam engine at the same address. Later that month, the *Sun* noted that the city council had accepted Dickinson's invitation to witness a test of the gun at the rear of 24 North Street on February 22. Another clipping from the beginning of April links the name David A. Dickinson to the same address.[26] Based on the involvement of the "brother" of Charles Dickinson in exhibits of the gun in Milwaukee, it is likely that David A. Dickinson was that brother, and that his involvement in marketing the gun accounts for the notices above and subsequent mentions of "W. D. A. Dickinson" in regard to the gun.[27]

They had no way of knowing it, but the decision Charles Dickinson and his associates made to take the gun to Baltimore would bring it national and international attention as events turned violent. That gun was similar in size to a steam fire engine, and like a fire engine was to be hauled into position by a team of horses. Also like a fire engine, it had a vertical boiler resting on a wheeled platform. A round, tub-shaped shield covered the barrel assembly to protect the operator from the machinery. A second curved iron shield covered the mechanism and would serve to shelter its operator from enemy fire.[28] Period press accounts say that the barrel rotated at about 1,600 times per minute and that it could throw three hundred two-ounce projectiles per minute a distance of less than a hundred yards. Projectiles left the barrel through a slit in the gun's shields.[29]

Although little remembered today, self-propelled steam devices like the "J. C. Cary," a fire engine from 1858, no doubt helped make the idea of a steam-driven and propelled gun seem realistic to the readers of Harper's Weekly. *(Engraving from* Harper's Weekly, *November 30, 1858/Library of Congress.)*

Dickinson made a number of strong claims for the gun in a written statement probably distributed to those attending his demonstrations and later picked up by the press. The gun was "a triumph of inventive genius, in the application and practical demonstration of centrifugal force . . . this most efficient engine stands without a parallel commanding wonder and admiration at the simplicity of its construction and the destructiveness of its effects; and is eventually destined to inaugurate a new era in the science of war." It was shot-proof, easy to move from place to place, and could be built in versions to fire missiles ranging in size from one ounce to twenty-four pounds, with a range and force similar to gunpowder weapons but—with the smaller rounds—at a rate of fire of between one hundred and five hundred per minute. Useful at sea and against infantry, it would "give the powers using it such decided advantages as will strike terror to the hearts of opposing forces, and render its possessors impregnable to armies provided with ordinary offensive weapons. Its efficiency will soon be practically demonstrated, and the day is not far distant when, through its instrumentality, the new era in the science of war being inaugurated, it will be generally adopted by the Powers of the Old and New Worlds, and, from its very destructiveness, will prove the means and medium of peace."[30]

The demonstrations and Dickinson's words seem to have attracted scant attention before April 19, 1861. Yet within days of that fateful date, the gun would be known across the United States.[31]

3

First Blood

On April 18, 1861, word reached Baltimore that the first Union troops answering President Lincoln's call for volunteers would soon arrive by rail. Noted Baltimore industrialist Ross Winans, who had already had some association with those advocating states' rights, was that day participating in a State Rights and Southern Rights Convention held at Taylor's Hall.[32] In fact he authored a resolution at the convention saying that Lincoln's intention to recapture forts in seceded states would lead to "sanguinary war, the dissolution of the Union, and the unreconcilable estrangement of the people of the South from the people of the North." His resolution also called the massing of free state militias in Washington a "standing menace to the State of Maryland, and an insult to her loyalty and good faith, and will, if persisted in, alienate her people from a government which thus attempts to overawe them by the presence of armed men, and treats them with contempt and distrust." It also suggested that all good citizens should "set aside party differences and present an unbroken front in the preservation and defense of our interests, our homes and our firesides, to avert the horrors of civil war, and to repel, if need be, any invader who may come to establish a military despotism over us." As it happened, this resolution was published on April 20, 1861, in a paper filled with news of the Baltimore Riot, a coincidence that no doubt helped establish Winans' reputation as a secessionist.[33]

About 2 P.M. on the eighteenth, a battery of artillery and a group of Pennsylvania recruits arrived at Bolton Station. "The regulars marched to Fort McHenry, and the volunteers went down Howard Street to Camden Station," the *Sun* reported. Several thousand jeering citizens followed the volunteers, who wound up at Mt. Clare, where a train had been assembled.

Nicholas Biddle, attached to the Pennsylvania recruits who arrived in Baltimore on April 18, was seriously wounded by a thrown brick as the men marched through the city to Mt. Clare. He was photographed after the war. (Photograph by W. R. Mortimer / Library of Congress.)

George Kane, marshal of police, and 120 of his men managed to keep order, and the troops made it safely out of town. But mobs and police remained in the streets. Clashes broke out between secessionists and unionists.[34]

The stage was set for the arrival, at about 11 A.M. on Friday, April 19, of the 6th Massachusetts Volunteer Militia and a larger body of unarmed Pennsylvania troops. These men would arrive at the President

As Union volunteers traveled through Baltimore in response to President Lincoln's call for volunteers, tension rose in Baltimore on April 18, and flared into violence on April 19, as the 6th Massachusetts Infantry clashed with southern sympathizers in what has become known at the Baltimore Riot. (Engraving from Leslie's Illustrated Newspaper, *April 30, 1861.)*

Street Station on the east side of the city, and because a municipal ordinance barred locomotives on the streets, individual cars had to be uncoupled and drawn by teams of horses from President Street west across the north side of the harbor to Camden Station. The distance was approximately a mile, and an angry mob of secessionists soon formed to block their way, eventually forcing several cars to turn back. Four companies of the 6th detrained, formed into column, and set out to cover the distance on foot. The mob fell upon the soldiers, some lunging into their ranks in attempts to steal their rifles. Shots were fired from the crowd, and stones rained down upon the troops from upper-story windows. In self-defense, the soldiers opened fire. When all was said and done, sixteen people—four soldiers and a dozen civilians—lay dead, and scores more had been injured.

Saturday, April 20, found secessionists engaged in a wild search for weapons. Warehouses, foundries, and gun and cutlery shops were broken into and any weapons found in them taken by force. Pro-union newspaper offices were attacked, and violence broke out against the homes and businesses of those known to hold union sympathies. Rumors of more troops approaching from Pennsylvania brought thousands of armed men into the streets. Church bells tolled, and fears of an impending attack by invading troops or from Fort McHenry added to the crisis.[35]

IE GUNPOWDER CREEK RAILROAD BRIDGE, ON THE PHILADELPHIA AND BALTIMORE RAILROAD, BY THE MARYLAND S]

In the wake of the riot, railroad bridges north of Baltimore were burned to ensure that additional federal troops could not travel to the city by rail. (Engraving from Leslie's Illustrated History of the Civil War, *1895.)*

Ross Winans: Pictured here late in life, in his prime, Ross Winans was a force to be reckoned with, particularly if one's patent came to close to one of his many railroad related inventions, or if one dared to question the merits of his locomotive designs. (Photograph courtesy of the B&O Railroad Museum.)

Ross Winans helped improve the design of the York, the B&O Railroad's first locomotive, purchased after design competition. (From Rambles in the Path of the Iron Horse, *1858.)*

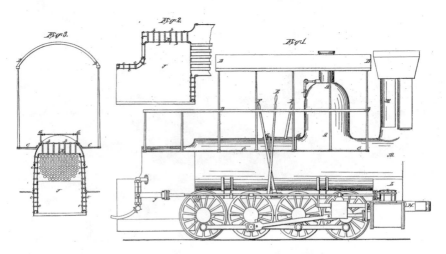

The Winans Camel, probably named for the hump-shaped steam dome atop its boiler, was slow but designed for the mountainous western portions of the B&O's line. (United States Patent, No. 19,962.)

⊰ 4 ⊱

From Railroad Pioneer
to "Traitor"

U NDERSTANDING THE EVENTS that unfolded in the wake of the riot requires a brief look at the life of Ross Winans. Born in New Jersey in 1796, Winans seemed destined for a life on the farm until he became interested in mechanical things and invented a plow and a method of fulling cloth by steam. He followed that with an improved type of wheel for railway cars. He licensed his invention to the fledgling Baltimore & Ohio Railroad and thus became involved in manufacturing its equipment prior to the road's construction.

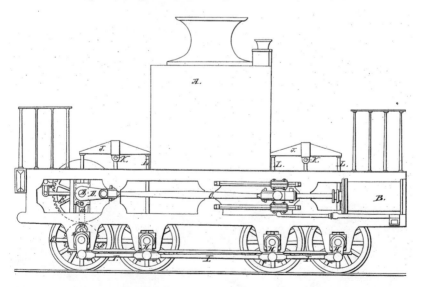

The Winans "Muddigger" was an important step up from the B&O's early engines with vertical pistons and walking beams. Its horizontal driving action helped apply more power to its driving wheels, increasing its ability to pull cargo, which was crucial to the road's growth. (United States Patent, No. 3,201.)

Winans also was among those to introduce steam engines on the B&O, and for a quarter of a century he worked to improve railroad technology, first in the employ of the B&O and later on his own. The culmination of his career was a coal-burning locomotive type called the "Camel" for the way its cab was perched atop its furnace. Rough and unconventional, Camels were nevertheless particularly good at one thing—muscling freight up and down the B&O's tracks in mountainous western Maryland and Virginia. His sons Thomas and William pursued their careers in Russia and made fortunes building and equipping the Moscow-to-St. Petersburg Railway.

The Winans Camel was unorthodox, slow, and roughly finished, but it could muscle heavy coal cars through the mountainous sections of the B&O's main line. In a time when most engines burned wood, it was among the first successful coal-burning engines. Though primarily used by the B&O, they were used by some other American railroads, and there is evidence that they were used in Russia by Winans' sons as well. (From the Collection of the B&O Railroad Museum.)

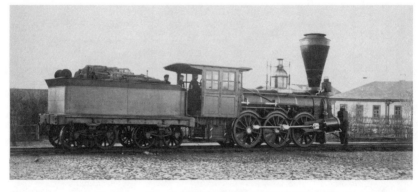

Locomotive on the Russian railroad. (Courtesy of DeGolyer Library, Southern Methodist University, Dallas, Texas, Image No. Ag1982.0226x.)

Thomas Winans made a substantial fortune building and supplying the Moscow-to-St. Petersburg Railroad for the Russian government. In 1861 he was working with his father on the Cigar Ship. (Courtesy of the Maryland Historical Society.)

By the spring of 1861, Ross Winans was known across the United States for aggressively defending railroad patents in court. Though by 1860 his work building locomotives had ended several years before in a falling out with the B&O over their design, he owned a large machine shop that was still in operation. The firm's order book for 1860 and 1861 shows regular requests for wheels, boilers, repairs to boilers, several large soup kettles, and sheet copper. Experiments with his imaginatively designed "Cigar Ship" were also ongoing and required the use of the larger foundry and machine shop at Mt Clare, as well as a smaller shop at his shipyard at Ferry Bar.[36]

After going out on his own, Ross Winans set up shop in a large factory complex where he built locomotives for the B&O and other railroads. (From Rambles in the Path of the Iron Horse, *1858.)*

Ross Winans' long and colorful career, his wealth and fame from his novel ship, his politics and close personal and financial ties to city government, and a simmering animosity in the northern press after years of lawsuits, placed him on a collision course with the federal government in the spring of 1861. Whether he simply wanted to help Baltimore authorities maintain order or intended to strike a blow against federal troops remains a mystery, but in the days after the riot his company's workmen set about repairing guns and making a variety of items for military use. He was, northern papers said, a traitor worthy of a traitor's fate.

As the B&O's operation grew, so did its shops at Mt. Clare, shown here in 1858. (From Rambles in the Path of the Iron Horse, *1858.)*

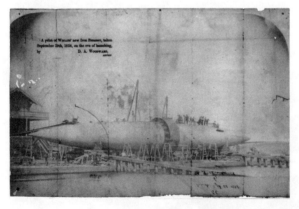

The Winans Cigar Steamer, built in 1858, was a test bed for what Ross, Thomas, and William Winans hoped would be a revolution in the passenger and shipping business. By April 1861 they were laying plans for the next generation of their ideas, which were interrupted by the family's association with events in Baltimore. They would continue their work with another Cigar Ship launched in England in 1864. The 1858 ship caused some concern at the beginning of the Civil War, but did not leave its dock. (Photograph by D. A. Woodward / Library of Congress.)

⊰ 5 ⊱

The Gun Takes the Stage

I N THE CONFUSION and general rush by municipal authorities to gather weapons after the riot, the steam gun was seized by the city, either from Dickinson or from his associates. One period press account suggests that a "Mr. Barker," probably the same man who had helped fund the gun's construction and was with it in Baltimore while it was being prepared for exhibition in Washington, was called home to attend to business. One period press account, probably supplied by attorney J. L. Rider, suggested that "In the absence of Mr. Barker, and without his knowledge, the authorities of the city of Baltimore ordered the gun put in order at once." Dickinson later claimed that "the traitor Marshal Kane seized the gun with the intention of using it to prevent the passage of loyal troops, through the city and took the inventor [Dickinson] prisoner." Whatever the case, the gun was now under the control of city authorities.[37]

The most well-known image of the gun depicts it outside what is presumably Ross Winans' locomotive factory in Baltimore. The identity of those shown is not clear. The man in the top hat may represent Winans, or he could represent the gun's real inventor and builder, Charles S. Dickinson. (Engraving from Scientific American, May, 25, 1861.)

Clues to what happened next are buried in a flurry of press accounts describing Baltimore's preparations for defense. Weapons were collected in a makeshift arsenal at the Old City Hall on Holliday Street. They included nine hundred surplus federal muskets from Colonel Isaac M. Deacon, 1,137 muskets and eighty rifles bought by city authorities from F. W. Bennett & Co., and 159 muskets from the Central Male High School. Additionally, according to the *Baltimore American and Commercial Advertiser* of April 22, "five field pieces of the students at St. Timothy's Hall College in Catonsville"—an Episcopal school for young gentlemen that included artillery training in its instruction—"were removed by the order of the authorities, and with the Patent Centrifugal steam gun of W. D. A. Dickinson, placed in Holliday Street in front of the Old City Hall."[38]

The *Sun* of the same day mentioned the St. Timothy's guns and added that, "The centrifugal steam gun of Mr. Dickinson, on exhibition in this city for some time past, has been purchased by the city, and will be used in its defense."[39] According to the *Sun*, Dickinson's gun was about the size of a steam fire engine and could fire three hundred projectiles per minute. The authorities intended to "plant the gun at the head of the street up which the invading troops attempt to march, and by signals to clear the streets of citizens and sweep the ranks."[40]

The following day the *Sun* carried its own report on the weapons from St. Timothy's Hall, to the effect that municipal authorities had requested the school's muskets and artillery pieces be turned over to them. When the city asked for the battery, "Mr. [Libertus] Van Bokkelen [St. Timothy's headmaster] immediately took steps to have it put in complete order, in which condition Col. [Isaac Ridgeway] Trimble . . . received it from Winans' shop."[41]

The *Sun* of April 23 also reported that "The Messrs. Winans are busy at the new foundry shops at Mount Clare casting cannon balls and grape [shot] in large quantities at the order of the city. Yesterday a steam gun from the machine shops of the Messrs. Winans was brought into Holliday Street, and exhibited for some time."[42] By "the Messrs. Winans," the paper meant Ross Winans and his son Thomas, who had recently returned to the United States after years in Russia building and operating the Moscow-to-St. Petersburg Railroad. Thomas was in partnership with his father (and with brother William, who was still in Russia looking after his own and Thomas's business interests) in the effort to develop the Cigar Ship and build a shipyard facility for it at the Ferry Bar.

The phrase "from the machine shops of Ross Winans," bears close attention. Taken alone, it has at times been considered proof of the elder Winans' participation in the gun's invention and construction. In reality, it is consistent with the firm's role as armorer for the Board of Police

Commissioners. With a large foundry, machine shop, and trained workmen, Winans & Company was well suited to undertake work that the city fathers desperately needed done in the days after the Baltimore riots, namely repairing cannons and making shot and pikes. Combining scattered news accounts reveals a pattern. Both the St. Timothy's battery and the Dickinson gun were secured from their respective owners by civil authorities. The St. Timothy's guns were said to have been repaired by the firm, while the steam gun, which was seized to "be put in order at once," was described as having come from their shop.

An April 22 letter to Winans & Company from Isaac Ridgeway Trimble, who was in charge of organizing citizens to defend the city, suggests why the gun would have been brought to the firm's shop prior to its public display. "Winans or any other gentleman who can prepare a prop[osal] to make balls for the Dickinson centrifugal gun is requested to execute that work for W. Dickinson," Trimble wrote.[43]

Trimble's note is important because it indicates that city authorities made the request on behalf of Dickinson, who would have received the proposal and thus been responsible for any work contracted under it. Surviving records of the city's account with Winans & Company do not contain entries for repairs to the St. Timothy's guns, which were probably paid for by the headmaster of the school. Three hundred pounds of "shot for the steam gun" were later found among city weapons in the custody of George P. Kane, marshal of police. However, given the press accounts noting the shot-making, and the surviving request for a proposal, it is likely that the shot came from Winans & Company, though it was not listed specifically as "steam gun shot" in the surviving account record.[44]

That Winans & Company was making munitions was widely known at the time, and Winans was and is often thought to have been doing so in support of the Confederacy. Surviving accounting records from the firm, together with munitions orders, prove that the work was in fact conducted at the request and expense of Baltimore's Board of Police Commissioners. The infamous Winans' Pikes were already being delivered to city authorities by April 21.[45]

Though he was sympathetic to the Southern cause, Marshal of Police George P. Kane deployed his men to help the 6th Massachusetts get safely out of Baltimore after its clash with angry pro-Southern citizens. (Courtesy of Daniel Toomey.)

Baltimore authorities turned to a time-tested anti-cavalry weapon, the pike, in the after-math of the riot. Ross Winans' factory turned out thousands of them at the request of city au-thorities, one the last appearances of a weapon long outmoded by rifled firearms. (Photograph courtesy of Daniel Toomey.)

Between April 22 and May 7, the company made and delivered 222 shells of canister, 206 of grapeshot, 99 cannon balls, 24 ladders, 10,327 pounds of lead musket balls, and 3,400 pikes for the police board. The sheer amount of matériel and the hundreds of hours spent in smelting, carpentry, and boiler-making, are sufficient to clearly demonstrate that Ross Winans, then in his mid-sixties, could not have made all those arms and ammunition himself, but that his company did so at the city's request.[46]

While the gun was being displayed on Holliday Street on April 22, 1861, Maryland's governor, Thomas Holliday Hicks, called a special session of the legislature to meet in Frederick, far from Annapolis, which was then occupied by Union troops. On the twenty-fourth, Ross Winans and nine other state's rights candidates were elected to represent the city in the special session, beginning April 26.

News accounts regarding the Winanses multiplied. Ross Winans' election on a state's rights ticket, his firm's involvement in making shot and pikes, news that a military unit bearing his family's name was to be found in Baltimore, the growing link to the steam gun, suggestions that Thomas Winans had pledged his entire $7 million fortune to aid the Southern cause, and reports that Ross was said to have bought seven thousand rifles for the city's defense put him on a course that could only end in confrontation with federal authorities.[47]

It is impossible to judge from available sources how long the gun was displayed on Holliday Street. It could have been anywhere from a few hours on April 22 to several days, as was later suggested by the *Baltimore American & Commercial Advertiser*: "During the dark week of April, it was taken possession of by our city authorities and kept, along with other heavy munitions, in front of City Hall." The paper also noted that "after the war"—by which it meant the April 19 riot—it was returned to its owner and inventor, and taken to Winans' shop for repair. This occurred at some point between April 22 and May 7, when an addition to the Board of Police Commissioners' account in the amount $606.82 for the repair was recorded.[48]

The repair raises the question of what had happened to the gun while it was in the city's custody. Had it been mishandled, or had its exhibition at the Old City Hall included a demonstration? Given that the gun's capabilities had been demonstrated publicly after its arrival in Baltimore, further demonstration would seem to have been unnecessary, but as more Federal troops were expected to arrive at any moment, it may simply have been kept fired up during the height of the excitement in case it was needed. While Winans & Co. was making munitions and repairing the gun, Union troops under General Benjamin F. Butler seized Relay, Maryland, the strategically important junction of the Washington and Frederick branches of the B&O, to stem the steady flow of men and goods headed toward the Confederates at Harpers Ferry.[49]

On April 27, city authorities asked Winans to stop making pikes. Company records indicate that the firm complied, although it did deliver an order of grapeshot on May 2.[50] Two days later the *New York Evening Post* incorrectly reported that "Winans is running 700 men night and day, in his immense establishment, casting cannon, shot, and shells, putting up grape and canister, and preparing munitions of war." That same day, the *New York Tribune* reported that "Among the men whose names should never be forgotten, until they have been duly punished for the atrocious crimes in which they have involved themselves at Baltimore, Ross Winans, Thomas Winans, Abel of the Baltimore Sun, [George] Kane, the Police Marshal, S. Teakle Wallis, and some others, are already known to the country. They are all traitors of the blackest dye, and merit the traitor's doom."[51]

On May 4, the *Tribune* published a description of the "Winans Gun," which a correspondent had seen at the Winans & Company shop. He described its boiler as being like that of a steam fire engine. "There is but one barrel, which is of steel, on a pivot, and otherwise is like an ordinary musket-barrel. It is fed or loaded through a hopper entering the barrel directly over the pivot. The barrel has a rotary motion, and performs the circumference, by machinery attached, at the rate of about sixteen hundred times a minute." It could throw three hundred two-ounce projectiles per minute, and the barrel revolved inside a drum of boiler iron. Its range was said to be "not over 100 yards at best." The gun weighed 6,700 pounds. The *Tribune's* editors added: "It is the opinion of our informant, that the gun does not warrant the expectations of the inventor, and that it is not likely to be of much service."[52]

The *Tribune* account was perhaps the first to label the machine the "Winans Gun." Early reports from the *Sun* and the *American* consistently linked it to Dickinson, while later reports from the *New York Tribune* and *New York Times* linked it to Winans. These and other Northern press accounts no doubt helped cement the perception that Winans was the gun's inventor and builder.[53]

Harpers Ferry, shown here in 1860, where the men who attempted to remove the gun from Baltimore hoped to sell it to Confederate forces. (Photograph by George Stacy, ca. 1860 / Library of Congress.)

Benjamin F. Butler's unauthorized action in seizing Baltimore prompted General Winfield Scott to reassign him, but he had helped ensure that Baltimore and Maryland remained in the Union. (Photograph by Brady Studios / Library of Congress.)

Colonel Edward F. Jones, commander of the 6th Massachusetts Infantry. (Courtesy of Daniel Toomey.)

⇥ 6 ⇤

Escape & Capture

Between May 7 and 9, 1861, the steam gun was removed from the Winans & Company shops to Charles Dickinson's yard.[54] "Monday of last week Dickinson told me the city didn't want the gun," John Bradford told General Butler during a subsequent interrogation on May 11. "I then proposed to Mr. Dickinson [that] the gun be taken to Harper's Ferry." Bradford enlisted two friends to assist in the endeavor. "Wednesday we went together to see [John A.] McGee [to find out] if it could be taken and what it might cost." They agreed that Bradford would pay fifty dollars then and another fifty dollars when they reached Harpers Ferry. Bradford intended to sell the gun to Virginia for "defense against invaders," specifically Federal troops.[55] During his interrogation, McGee told Butler: "We went down to Dickinson's yard yesterday morning [May 10] – took it up to my place where it stood until about 7 o'clock – about 12½ o'clock we got to Ellicott's Mills, about 9½ miles."[56]

John McGee and Richard Harden and their team hauled the gun away from McGee's yard, while Bradford and Dickinson accompanied them in a buggy.[57] They made no attempt to conceal the gun, because they reasoned that if they were stopped by Federal troops the gun would be considered private property, since Virginia and the federal government were not at war.[58]

Meanwhile, when Colonel Edward F. Jones, of the 6th Massachusetts, learned the gun was leaving Baltimore, he and Colonel George Lyons, in command at Relay, dispatched a company of the 8th New York Militia, one from the 6th Massachusetts, and two artillery pieces from Cook's Battery, Massachusetts Light Artillery, in pursuit.[59] About 11 A.M. on the tenth, there was a burst of activity at Relay as the troops prepared to commandeer the noon train for Ellicott City.[60] In Washington, four companies of the 8th Massachusetts turned out to march to Ellicott City at a moment's notice, in case the gun's capture provoked pro-secession forces to violence.[61]

While the troops at Relay were forming and boarding the train they had commandeered, Robert Hare, Esq., a local resident who had been serving as a courier for Butler since the general's arrival in the area, rode ahead alone on horseback and overtook the gun. "The person having it in charge, at the moment of the capture, was not present," wrote a member of 6th Massachusetts. "Captain Hare stepped up to the driver and said to him – 'Halt! Get off this animal!'"

"Get off? What Fur?" quietly asked Mr. Muleteer.

"Because I command you. Dismount instantly!" and a wicked looking revolver was placed within an inch of the muleteer's head. The man came down with a jump.

"You are my prisoner! Move at your peril!" And turning to the man on the gun Capt. H. gave similar orders and a similar caution. The men were unarmed, and the Captain proceeded with them to the Mills, and procured the assistance of three millers, and with them returned to the gun, and the "institution" was put in motion. They had proceeded but a short distance when an individual rode up at a furious rate.

"Who stops my Team?" he exclaimed.

"Is this your property?" returned the Captain, with a Yankee answer.

"Yes; it is mine; or rather I am the agent for it."

"Then, in the name of the United States of America, I make you my prisoner!"

The "agent" placed his hand on his breast, as if to grasp his revolver; but Capt. Hare was on his guard, and his revolver covered the man's temple.

"Will you dismount!" it was a civil request, but it was most urgent. The man dismounted . . . the men then gave their names – John McGee, Richard E. Hardenge, and J. Stucker Bradford, (the "agent"). The muleteers say they are good "Union Men," that they simply made a contract to furnish four mules and two men for two days' service, and they did not know the nature of the service required. It will be bad business for them. The gun and the prisoners were brought here [Relay]; the gun was conveyed to the camp, and the prisoners sent to Annapolis.

No mention of Dickinson is made here, though both local and New York papers reported his alleged capture.[62] Returning to McGee's account: "Capt. Hare rode up and asked me to surrender my wagon. Bradford and Dickinson were ahead." Sadly, the second page of McGee's account, which might have shed light on Dickinson's role in the events, has been lost.[63]

After capturing the men who had attempted to convey the gun to Harpers Ferry, Butler's adjutant marched them on foot to Ellicott's Mills (now Elliott City) where he secured the assistance of several local men in dealing with the prisoners. Meanwhile, a detachment of troops arrived, secured the gun and took it back to their camp at Relay. (From Rambles In the Path of the Iron Horse, *1858.)*

"Upon learning of the gun being seized (we being in advance of it on the road) I immediately returned to it demanding of the officer in charge by what right it had been seized + and was therefore arrested," Bradford recounted. He also noted, "There was no portion or part of the gun taken off for the purpose of disabling it. Was put into our carriage – it was intended but was not done – The bullets were put into our carriage. I think there was about 200 of th[em]."[64]

"The Relay camp was quite lively over the capture last evening [May 10] and the gun is the subject of much curious inspection," the *Sun* reported on the eleventh. "It has been erroneously asserted that the centrifugal gun belongs to the city of Baltimore. This is not so. The gun belongs to the inventor, and if there existed any intention to carry the gun out of the city it was not known to the city authorities." The *Sun* also noted that Dickinson had been arrested with the gun.[65]

An unruly party of visitors arrived at Relay on Saturday, May 11. Some came to present a flag to an infantry regiment, perhaps the 6th Massachusetts, but others came with "the expectation of witnessing the experiments with the steam gun which Gen. Butler's command captured with so much 'coolness, promptitude and zeal,'" wrote the *Daily Exchange.* "But they were doomed to disappointment on account of the troops not having captured all of the gun.

They have just enough to render the gun useless, as they have the steam engine and 'running gear,' and somebody else has the gun. As much as they have of it is located in the Massachusetts camp, where the curious can inspect it and 'make the most of it.'"[66] The *Sun* provided a similar account: "The steam gun, that all had come to look on as a death dealing engine, stood as harmless as an old barn fan. It now turns out, that the inventor, Mr. Dickinson, who was not taken with the gun, had with him in a buggy, a short distance off, all the important parts of the machinery used in the working of the gun, and escaped with them by driving rapidly away." According to the *Sun*, the men captured with the gun had been sent to Annapolis, and Dickinson was also there.[67]

The discrepancy between these accounts—one from a participant saying that the gun was intact at its capture, the *Sun* and *Daily Exchange* detailing its uselessness—is perplexing. Reports of Dickinson's whereabouts immediately after the gun's capture are likewise confusing. Some say that he was captured by Federal troops and taken to Annapolis, but no record of an interrogation has been located alongside those of the other teamsters captured with the gun, and published accounts of their interrogation do not mention him. According to another press account, he was not captured but was in Annapolis.[68]

On the evening of Monday, May 13, and into the early hours of Tuesday, May 14, Butler's troops, under the cover of a rainstorm, entered Baltimore and encamped on Federal Hill. Already anxious citizens awoke to Union soldiers in their midst. Importantly, the *Baltimore American* of the fourteenth provides an opportunity to compare the notes of the interviews taken by Butler's clerk, mentioned earlier, with the published version of events. According to the paper "Yesterday morning the three . . . custodians of the steam gun, were brought before General Butler, at Annapolis, for examination. The two teamsters told so straightforward a story that the General at once released them, and returned them their mules. John Bradford, the chief of the party, fared differently." When Bradford asked for a lawyer, Butler told him, "an honest man would not want counsel." The examination would decide if Bradford should be held for trial. If he gave no statement, Butler said, he would be taken into custody.

Bradford explained that he was a native of Baltimore but now lived in Virginia and considered himself a Virginian. "As he [Bradford] is informed and believes, the steam gun was seized by the city authorities of Baltimore on the day succeeding the attack on the Massachusetts Sixth Regiment by the mob. About a week since one Dixon, or Nickerson . . . said to the prisoner that the city did not want the gun and did not intend to keep it." Bradford suggested taking it to Harpers Ferry to sell. "Previous to this time the twain had had some conversation about the sale of the gun. Nickerson, or whatever his name is, hung fire at first, but finally assented to the proposition."

During the Federal occupation of Baltimore, Fort Federal Hill provided a commanding vantage point for artillery and for observing events in the city. (Engraving by E. Sachse & Co. Ca. 1872 / Library of Congress.)

Bradford told Butler he had paid several teamsters fifty dollars down and promised another fifty dollars when they reached Harpers Ferry. He had borrowed the money, but refused to say from whom. Butler asked Bradford if he knew of Virginia's ordinance of secession. Bradford did, but "did not believe the people would sustain it. He intended or expected that the gun should be used to defend Virginia against invasion. 'Invasion by Federal troops,' suggested the General. 'No,' replied Bradford, 'I don't see it in that light.'" Butler lectured him about "shuffling and finally Bradford acknowledged that he meant an invasion by Federal troops, if the General pleased to call his forces by that name." Bradford said he had acted alone, and gave a speech about secession, states' rights, and the "inviolability of state soil," which, Butler said, "the Secretary would not take the trouble to preserve. He was forthwith sent to jail."

According to the *American and Commercial Advertiser,* "Dickerson" had patented the gun and Bradford's only interest in the matter was if it were sold. "He believed that when in order it would be a formidable weapon. He had seen experiments with it." The story of the gun being taken out for long-range trials amused Bradford. He had not heard of that until Butler told him.[69]

Also on May 13, readers of the *Daily Dispatch* in Richmond, found the following item regarding the gun.

Thomas Winans, of Baltimore, has written a letter denying the statement that he had ever tendered the half million loan reported to the Marylanders; and declaring that the assertion in regard to manufacturing steam cannon for them is equally unfounded. With regard to the casting of cannon, balls and grape-shot, he adds the assertion is so far true that the same establishment was, with others in Baltimore, employed by the city authorities to furnish these articles, when they believed that the peace of the community required military aid. For this purpose the authorities made an appropriation of half a million dollars.[70]

Meanwhile, in Pennsylvania, readers of the *Philadelphia Inquirer* learned that "Orders were issued . . . to all commissioned officers, to arrest Ross Winans, whose steam gun has been captured. He will be tried for treason – first, to find whether he was accessory to sending the gun to Harper's Ferry, next, whether he has aided, by means of money or otherwise, the traitors. If caught, he will attempt escape on the plea that the gun was sold to the City of Baltimore, which was undoubtedly the case, but done at a time when the city was in state of rebellion."[71]

While a denial from the family and word of Winans' impending arrest were circulating in other parts of the country, Federal troops were searching for Winans closer to home. They found him on May 14.

"As members of the Legislature, which adjourned yesterday [May 14], were returning to Baltimore in an extra evening train, it stopped a few minutes at the Relay House when an officer entered the cars, and approached Ross Winans, Esq., one of the members of the House of Delegates from Baltimore City, and courteously inquired if he was Mr. Winans. Receiving an affirmative reply, the officer said he wished to speak with him. Winans asked him what it was, and the officer said he had an order for his arrest." Winans was escorted into the depot. Members of the legislature tried to follow, but only Governor Thomas Holliday Hicks was allowed inside. Hicks returned in a few minutes, not having "been given satisfaction" in regard to the cause of the arrest. Hicks's offer to guarantee Winans' appearance was refused, but he was assured that Winans would be comfortably cared for until he could be "examined" by federal authorities. Winans was taken to Annapolis, then to Fort McHenry.[72]

"Ross Winans, Esq., the secessionist, was arrested at this place . . . the arrest was made in the cars, about eight o'clock in the evening, by Col. Fay, an aide of Gen. Butler, Lieut. Moses Emory, of the Washington Light Guards, and your correspondent, as Orderly," W. J. D., a member of the

Following the crisis in April, Maryland's governor, Thomas Holliday Hicks, called a special session of the Maryland legislature to meet in Frederick. He was present when Federal troops seized Ross Winans as his train stopped at Relay on the way back from that session. Hicks tried to learn the reasons behind Winans' arrest but was given "no satisfaction" by the arresting troops. (Hicks: Brady Handy Collection / Library of Congress; troops at relay: courtesy of Daniel Toomey.)

6th Massachusetts, wrote to his hometown newspaper from Relay. "It was done in a quiet way, but it was the occasion of bringing the whole camp to arms. Mr. Winans is a gentleman and a Christian, and right in every [way] saving politics. He is a noble-looking man, with a massive gray head and beard. During his conveyance to Annapolis, he entertained the guard with anecdotes of travel and interesting conversation "[73]

Winans was captured quietly, but the evening was not so quiet for the 8th Massachusetts nearby. "About 11 o'clock in the night our camp was thrown into a wild state of excitement by the firing of our outposts, or 'picket guard.' Some thirty to fifty shots were fired, and the 'long roll' of the drum called us from our slumbers. The lines were quickly formed, and the order given to march by the 'right flank,' file left. We marched up a steep hill, covered by a growth of trees," wrote W. A. F., a member of the regiment. They expected an attack from the direction of Harpers Ferry. "After getting the boys in good order for the brush, a messenger arrived, and made the statement that the alarm was caused the arrest of Ross Winans."[74] A member of the 8th New York wrote an equally colorful account for the folks back home, saying that "Last night we captured at the Relay House Ross Winans himself. The mob attempted to rescue him: but the Sixth, of Massachusetts, came across the bridge like a streak of lightning to our rescue, and, after a little squabble, they retreated, leaving our prisoner in our hands."[75]

The reasons leading to the arrest included treason, according to the May 15 *Baltimore American and Commercial Advertiser*, and "selling the steam gun to the Virginians," according to the May 16 *Daily Dispatch* in Richmond, but the arrest was nevertheless confusing to citizens of Baltimore.[76] "There is considerable feeling in the city this morning in consequence of this arrest," a correspondent to the *New York Times* wrote from Baltimore on May 16. "Winans is a wealthy man, has employed a great many workmen in Baltimore," he added. "Why he has been arrested remains a mystery to the citizens here."[77] The correspondent believed that Butler had evidence that Winans "had recently visited Harpers Ferry for disloyal purposes and that he has manufactured a large quantity of cannon balls and other war materials for the enemy; that being unable to transport said war materials across the line of the Federal troops he proceeded to melt the balls in bars, thus destroying their identity; and that he had otherwise employed his means to aid and comfort those known to be in rebellion with the Federal authorities." He continued that Winans would be released if Butler found the charges incorrect, and noted that the steam gun was "neither invented nor built by Ross Winans" but that its inventor, "Mr. Charles Dickinson of Connecticut," had taken away a key piece on the way to Harpers Ferry, leaving it "useless."[78]

Winans' "visit" to Harpers Ferry is mentioned in "Affairs at Harper's Ferry," in the *Daily Dispatch*, on May 17, 1861. "Yesterday about forty of the members of the Maryland Legislature visited this place. The special object of this visit is not precisely known. Many supposed that they came here to protest against the seizure of the Maryland heights by the Virginia troops, but the political complexion of the delegation forbids such an inference. It may be worthy of notice that Ross Winans accompanied the delegation.—He believes that there will be no war of any consequence."[79]

The *Scientific American* would later say the charge was "that he had given aid to Virginia, by furnishing grape shot, &c. The utter fallacy of this was at once made apparent to the President, when he directed General Scott to issue an order for his unconditional release. The order was immediately sent to Fort McHenry, where Mr. Winans was under arrest, and he was at once released and returned to his family."[80] To the chagrin of General Butler, Baltimore attorney Reverdy Johnson, whom Butler would later describe as a "rank and bitter secessionist," went to Washington and secured Winans' release. "How much of Winans' $15,000,000 it cost him, I do not know, but it should have been a very large sum, because he relied on its potency," Butler commented.[81] Recent research reveals that Johnson took up the matter directly with Lincoln, who passed the matter to Secretary of State William H. Seward, who ordered the release. Butler,

who had planned to hang Winans in Baltimore's Monument Square as an example of what would happen to traitors, was not pleased.[82]

Years later in his autobiography, Butler fumed that Winans' wealth, his manufacture of five thousand pikes "to oppose the march of the United States troops," and a desire to make him serve as an example for Marylanders were the reasons behind his desire to hang him. Although he wrongly identified Winans as the builder of the gun, most of his anger seems to have been directed at the pikes, which he said had been used against his men during the riot. He had intended to form a military commission from the men of the 6th Massachusetts, who would of course recommend that Winans be hanged, and to then carry out the sentence. If a prominent and wealthy man such as Winans were executed it would have convinced Marylanders "that the expedition we were on was no picnic expedition."[83]

Baltimore attorney Reverdy Johnson rushed to Washington seeking Ross Winans' release after he was arrested at Relay while returning home from the state legislature's session in Frederick. (Photograph by Mathew Brady / Library of Congress.)

Famous for its part in the Battle of Baltimore during the War of 1812, which inspired the national anthem, Fort McHenry remained in Federal hands during the tumultuous days following the riot. Ross Winans would be confined there after his arrest. (Engraving from Harper's Weekly, July 13, 1861 / Library of Congress.)

THE WINANS STEAM GUN, CAPTURED BY GENERAL BUTLER'S COMMAND NEAR THE RELAY HOUSE, MD.

Leslie's Illustrated Newspaper *depicted the gun, now in the hands of Union troops, ready for action in defense of the strategic junction at Relay, Maryland. The barrel projecting outside the gun's iron shield is a bit of artistic license to make it appear more menacing—the real barrel operated under the cover of the shield, and a tub-shaped shield protected the gun's operator from any misdirected projectiles. (1861 engraving reprinted in* Leslie's Illustrated History of the Civil War, *1895.)*

7

Trophy of War

WHILE WINANS RETURNED to his home and family in Baltimore, the gun remained secure in the Union camp at Relay. Many accounts suggest that Dickinson removed key parts of it when escaping from Federal troops. This idea is one element of the story that is so frequently repeated that only those accounts presenting a different idea bear further notice here.[84] In addition to Bradford's testimony that the gun was intact upon capture, four additional accounts support his account. On May 15, General Butler reported to the War Department: "I also have the honor to communicate the capture of the steam gun, and the fact that I have found men in the 6th Massachusetts Regiment who have been able to put it in operation, and it is now in full working order."[85]

The next day a member of the Salem Zouaves wrote home noting that, "on a hill along side of us is encamped the 6th [Massachusetts] Reg. They have with them the famous steam gun which they captured on the road to Harper's Ferry. It is a curious looking machine, and they say it can throw 300 balls a minute, the distance of one mile, with the accuracy of a rifle. They have fixed it so that it works well, and if occasion offers it may work against those whom it is intended to serve."[86]

On May 25, 1861, the *Wisconsin Patriot* treated readers to a description of the gun, courtesy of the *Washington Star.*

> The steam gun . . . is one of the lions of the Massachusetts camp. It is an odd looking concern, bearing not a single indication to the un-practised eye of its murderous purpose. Through the intelligent aid of Captain Pickering, who seems to know what's what about most sorts of machinery, we obtained some sort of a notion how the thing

was worked. The whole concern which weighs perhaps five tons, is mounted on wheels. Externally, it has the appearance of a small two-horse power engine at one end, and the other, runs off into a sharp nose, not unlike an end of the Winans cigar steamer. This nose, however, which is merely the sheath to protect the machine and its operatives, is constructed of one and a quarter inch iron; and the expectation of the inventor was, apparently, that balls aimed at it would glance off harmlessly. In the opinion of those conversant with such matters a Minie ball would penetrate this sheath, while a 6 pound shot would of course knock the whole thing into a cocked hat. This pointed sheath or covering is divided nearly its whole length by a slit of three inches in width, affording an opening for the discharge of the gun. With this mouthlike slit dividing the sheath into ponderous jaws, and stretching from ear to ear, the affair has the look of some devilish shark nosed sea monster. Peering in at this opening not much is to be seen beyond a few cog wheels and a bit of mild looking cylinder, which, however is the mouthpiece of the centrifugal wheel, which, revolving at the tremendous rate of 350 times per minute, flings out a three ounce ball at each revolution.

The Massachusetts folks think that the machine does not amount to much, its unwieldiness being a fatal objection. If placed to command a narrow passage it might, however, do good service. Captain Pickering[87] says they will give it a trial today, anyhow to test its merits.[88]

Long after the war, Corporal James Whitaker of the 6th Massachusetts would tell the *Boston Journal* about this trial. A call was made for volunteers to try to operate it. Former mill machinists in the unit set to work on it, and it was placed on the hill behind the Relay House, a hotel near their camp, and fired. "A target about ten feet high and twenty feet long was set up a few hundred feet away, and after the machinists had got up steam and were ready to test it, they were told to go ahead and fire." Whitaker and the onlookers were astonished. "The rain of bullets about the size of an English walnut cut the target completely in twain, and it was evident that is would do great execution in cutting down a regiment of men," he recalled. "But it was also evident that it would be of little value in real warfare since it was unarmed and a single shot might put it out of commission, and it was abandoned as of no further use."[89]

Although the gun apparently remained at Relay, its story continued to grace national publications. On May 18, *Leslie's Illustrated Newspaper* published an engraving entitled, "The Winans Steam-Gun, Lately Captured on Its Way to Harper's Ferry." The gun was depicted stationed by the railroad tracks at Relay. A man in uniform is on the gun while two officers and a

Corporal James Whitaker of the 6th Massachusetts, in a photograph from the Boston Journal, *April 12, 1911, when he came forward to provide details of the gun after its capture.*

Corp. James Whittaker,
Company L, Sixth Massachusetts Volunteers, who lives in Stoneham.

sentry confer in the background. The artist took a bit of license by adding a fanciful barrel projecting from its iron shield, when in fact the gun's barrel revolved inside the shield. One additional point about the engraving is that it shows a team of horses hitched to a tongue attached to the gun's rear axle assembly, suggesting that horsepower was its intended mode of transportation rather than self-propulsion, as was sometimes suggested. Further, no cylinders, cranks or other driving gear are shown attached to the gun's front wheels.[90]

The War Department, meanwhile, was laying plans for it. On May 20, Butler was told the gun was "put at his disposal for transportation to a competent machine shop for tests & examination."[91] That same day Butler wrote to Governor John A. Andrew of Massachusetts, saying that the War Department had given him "leave to send home the steam gun captured by Col. Jones's men, for examination & tests of its practical usefulness, & I have no doubt the mechanics of Mass. will be able to find out its value, if it has any, or to improve if it is capable of improvement."[92] That same day, a small news item in the *Boston Transcript*, copying from the *Philadelphia Press*, noted that "The steam gun captured by Gen. Butler was not only made in Boston, but Dickinson, who had charge of the gun when captured, is a Boston man, and had it made under his own superintendence."[93]

The gun took an indirect route to Massachusetts, followed at every step by controversy concerning its effectiveness. On May 22, Corporal Fred Smith, then at Fort McHenry, noted in his diary that the gun had left for Washington, accompanied by a guard from the 6th Massachusetts. "The gun is generally looked upon as a destructive affair," he observed, "but the difficulty of ranging it, will, I think, deprive its possessors of using it to advantage."[94] The next day, Major W. H. Clemence, Butler's aide-de-camp, wrote the general to complain that "The pleasure of taking the steam gun to Massachusetts seems to be denied me." He had been ordered to return to Massachusetts, and had received an order from General George Cadwallader that it be held at Annapolis until further notice.[95]

May 25 found the steam gun featured in *Harper's Weekly.* "We herewith illustrate, from a photograph by Weaver, the celebrated steam gun, patented by Mr. Dickinson, and made by Mr. Winans of Baltimore. This gun was seized by Colonel Jones, of the Massachusetts Volunteers, when on its way from Baltimore to the Rebel Camp at Harpers Ferry, and is now used in protecting the viaduct at the Washington Junction on the Baltimore Branch of the Baltimore and Ohio Railroad. . . . The merits of the steam gun are a matter of some controversy," *Harper's* went on. "We shall probably know ere long what it can do."

A lengthy statement from Dickinson called the gun "a triumph of inventive genius, in the application and practical demonstration of centrifugal force . . . this most efficient engine stands without a parallel commanding wonder and admiration at the simplicity of its construction and the destructiveness of its effects; and is eventually destined to inaugurate a new era in the science of war." It was shot-proof, easy to move from place to place, and could be built in versions to fire shot from one ounce to twenty-four pounds, with similar range and force to gunpowder weapons, at from 100 to 500 times per minute. Useful on sea and against infantry, it would "give the powers using it such decided advantages as will strike terror to the hearts of opposing forces, and render its possessors impregnable to armies provided with ordinary offensive weapons. Its efficiency will soon be practically demonstrated, and the day is not far distant when, through its instrumentality, the new era in the science of war being inaugurated, it will be generally adopted by the Powers of the Old and New Worlds, and, from its very destructiveness, will prove the means and medium of peace." The *Harper's* engraving differed from that published in *Leslie's* in that it showed the gun in front of a factory, presumably the Winans & Company shop. Though it is hard to spot, the end of a tongue is shown attached to the rear axle, suggesting that rather than self-propulsion, it was intended to be horse-drawn. A man stands on the gun's platform with his hand in the chute where balls would be fed to its barrel. A man in a top hat stands in front of the gun. Several different versions of this engraving have been made over the years, with different numbers of people, as well as different shapes given to the protective cover over the barrel. This engraving, said to be made from a photograph, is perhaps the closest we will come to knowing what the gun looked like, unless "Weaver's" photograph yet survives.[96]

Scientific American also featured the gun on May 25, giving its front page to it and similar inventions from the past. Although its coverage closely followed that of *Harper's Weekly,* it provided additional details. "The annexed engraving represents a perspective view, taken from a photograph, of the famous steam battery, about which so much has been said within a few weeks, as being in process of construction by the Messrs. Winans of Baltimore," wrote the editors. "From a letter by Mr. Thos. Winans, published in the Baltimore papers, it appears that the machine belongs to the city of Baltimore, and that the only ground for connecting the name of the Winanses with it is the fact that it was sent to their shop for repair."[97]

Scientific American also presented Dickinson's claims with a clear description of the gun's mechanism, informed heavily by Dickinson's 1859 patent, and a woodcut of the gun's barrel. "A steel gun barrel, bent at an elbow . . . is caused to revolve by steam power with great velocity; when the balls being fed into the perpendicular portion, which is at the center of revolution, are thrown out of the horizontal arm by centrifugal force. A gate . . . keeps the balls from flying out until the barrel is in the desired position, when this gate is opened by the action of the lever . . . and the balls permitted to escape. To make sure against accident from the chance issuing of balls when the barrel is not in the proper position, a strong wrought iron casing surrounds the gun, with a slit, in one side through which the balls may pass. . . . Our cut represents the balls as being fed in singly by hand, but in action it is proposed to feed them in with a shovel." The article is of interest not only for the detail, but also for an imprecise quote. "Mr. Winans says that the shot from this gun will cut off a nine inch scantling at the distance of half a mile." Unfortunately, the journal did not clarify which Mr. Winans—Ross or his son Thomas—made that statement. The engraving showed the gun parked in front of their shop in Baltimore.[98]

Meanwhile, an account written from Annapolis on May 27 and subsequently published in the Richmond *Daily Dispatch*, indicated that the gun had gone not to Washington but to Annapolis, as planned. "The steam gun is now here, and an object of great curiosity to the Yankees. They flock around it every day, to study its points and seek, if possible, the problem which its inventor kept secret by preserving its most valuable parts." The report went on to say that "It is now as useless as so much iron, but I understand that Butler will shortly dispatch it to Lowell, Mass., his home, to be examined by some of the mechanics of that city. The papers are meanwhile fighting over the credit claimed by different detachments for its capture. Some two hundred men and a battery were engaged in the laudable enterprise, and the result of the exploit was a man, a boy, and a pair of mules."[99]

The June 1, 1861, issue of *Scientific American* frowned on the idea of steam guns in general and the Baltimore gun in particular. "In reference to this uncouth war engine, illustrated in our last number, one of the men who was arrested while in the act of conveying it to the secession camp at Harper's Ferry, says: — 'It requires fifty men to work it; it shoots behind and before and all around; it will certainly kill the fifty men employed in working it, besides dealing out its death strokes to thousands of those in whose defense it is employed.' A very amiable machine, indeed, for killing friend and foe. We suppose the inventor intended to use it for the purpose of committing suicide, or did he meditate the destruction of Ross and Thomas Winans?"[100]

A letter in the *Cambridge[Massachusetts] Chronicle*, dated May 28, 1861, and published on June 1 reported incorrectly that the gun had been taken to Annapolis, then to New York, and finally to Boston. In fact it did not reach Boston until much later.[101]

On June 3, while the gun was still in Annapolis, Dickinson's disgruntled former associate, William Joslin, wrote to General Butler, repeating statements he had made in the pages of the *Cleveland Leader*. "I wish to inform you that I am the inventor and patentee of the centrifugal gun taken on the way to Harper's Ferry by your army." Dickinson had taken advantage of him, "an inventor without money to protect myself and procured a patent on what he calls an improvement which is an absolute damage to my invention to defraud me and deceive the public." Joslin hoped Butler would allow him to work on the gun and put it in working order and said he could make it useful. He added that the gun had been built in Boston without his permission.[102]

On June 11, General Nathaniel Banks, then in command of the Eighth Army Corps headquartered at Baltimore, ordered Colonel Abel Smith of the 13th New York to send a "subaltern officer with a detachment" to escort the gun to Fortress Monroe. Smith had apparently been communicating with Banks regarding the seizure of "arms of ammunition at Charleston and on the Pautusant [Patuxent]," for Banks's adjutant, Robert Williams, replied: "This detachment will be able to perform such duties at the points on the Bay to which you refer, as your information may render necessary."[103] A few days later, on June 15, Private Phillip Woodruff Holmes of the 13th New York noted in his diary that "We are off at last, first to [Fort] Monroe to carry to that fortress Winan's steam gun." Another order from Banks, this one to Colonel J. S. Pinckney, 6th New York, on May 24, directed that the gun be sent, with one "subaltern and twenty men," to Fortress Monroe.[104]

"Nothing can be more absurd than the attempts . . . to apply steam directly and indirectly in projecting bullets — large and small — for purposes of warfare," opined *Scientific American* on June 15, adding that steam and centrifugal guns were too complicated. "The difference between the instantaneous combustion of the powder and the slow combustion of the coal will convince any person how superior the former is to the latter as a force for projecting missiles of war."[105]

The *Illustrated London News* also mentioned the gun in its June 22, 1861, edition. It was said to "have just been made by Mr. Winans, of Baltimore," and seized by "Colonel Jones, of the Massachusetts Volunteers when on its way from Baltimore to the Secessionist Camp at Harper's Ferry, and has since been used in protecting the viaduct at the Washington Junction on the Baltimore branch of Baltimore and Ohio Railroad." The article then presented Dickinson's claims for the gun that had appeared in *Scientific American* and elsewhere in the American press.[106]

Back in Baltimore, Federal troops were about consolidating their hold on the city—securing weapons and arresting officials friendly to the Southern cause. Early on the morning of June 27, General Banks dispatched a thousand troops from Fort McHenry to the home of Marshal of Police George P. Kane, took him in charge, and returned to the fort. Later that day, the operations of the Board of Police Commissioners were suspended and Colonel J. R. Kenly of the 1st Maryland Infantry was made provost marshal of the city.[107]

Once Kane was under arrest, troops still on the hunt for evidence against Winans, among others, searched the Old City Hall as well as the Central Police Station. A considerable amount of weapons and ammunition was found, including more than 80,000 Minié cartridges, 700 muskets and

As Federal troops searched the Old City Hall in Baltimore, which served as headquarters for Baltimore police, they found a variety of weapons, including projectiles for the steam gun, probably manufactured by Ross Winans' firm at the request of Isaac Ridgeway Trimble. (Engraving from Leslie's Illustrated History of the Civil War, *1895.)*

Union military authorities pursued a strategy of arresting city leaders sympathetic to states' rights or secession as a means of ensuring that they did not take Baltimore and by extension, Maryland, out of the Union. Marshal of Police George P. Kane was among those arrested on June 27, 1861. (Engraving from Leslie's Illustrated Newspaper, *July 6, 1861)*

rifles, a six-pound field piece and equipment, and shot including a cart load of canister shot from Old City Hall. The search of the Central Police Station at Holliday and Saratoga revealed 30 colt revolvers, 2 six-pound iron guns, 2 four-pound iron guns, half a ton of assorted shot, half a keg of shot for the steam gun, 120 flint muskets, 2 Hall's carbines, 46 rifles, 3 double-barreled shotguns, 8 single-barreled shotguns, 9 horse pistols, 65 small pistols, 112 bullet molds, 4cwt balls, and 8 dirk knives.[108]

About the same time, on the evening of June 29, the steam gun arrived at Fortress Monroe, according to an Associated Press dispatch.[109]

As the gun was completing the first leg of its journey north, William Joslin was pushing his claim to be the gun's originator, this time through the *Scientific American*. "Allow me . . . to set the public right, through your paper, as to the centrifugal gun, called the 'Winans' or 'Dickinson' gun. I am its inventor," he wrote in a letter in the July 6, 1861, issue. "Dickinson . . . derived the invention from me. I can, without fear of failure, build a machine that

After its stay in Annapolis, the steam gun was shipped first to Fortress Monroe, then north to Lowell, Massachusetts. (Engraving by E. Sachse & Co, 1862 / Library of Congress.)

will do more execution with . . . twenty men, without steam, than a thousand men can do with rifles or muskets. . . . After . . . my patent, Dickinson obtained a patent for what he calls an improvement which is an absolute damage to the machine, and balls can never be discharged correctly with it."[110]

On August 5, in a report of affairs from Fortress Monroe, the *New York Times* noted that the "celebrated Winans steam-gun" would be taken to Boston aboard the *S. R. Spalding*.[111] The ship reached Boston on the eighth.[112]

"The famous Steam Gun captured in Maryland by a detachment of the Sixth [Massachusetts] Regiment . . . and about which so much has been said and written, arrived in this city yesterday afternoon," the *Lowell Daily Citizen and News* announced on August 12. The paper noted that Dickinson was "formerly well known in this city as a teacher of dancing." After briefly giving some of the familiar claims made for the gun, the paper gave a detailed account of its particulars.

> The machine is a clumsy looking and poorly constructed piece of workmanship, weighing about two tons, and very much resembles a railroad snow plough. It consists of a steel gun barrel, bent at an elbow, which is made to revolve by steam power with great velocity; when the balls, being put in the perpendicular portion, which is at the centre of revolution, are thrown out of the horizontal arm by centrifugal force, a gate keeping the balls from flying out until the barrel is in the desired position, when it is opened by the action of a lever, and the balls permitted to escape. To make sure against accident from the chance issu-

ing of balls when the barrel is not in the proper position, a strong wrought iron casing surrounds the gun, with a slit in one side, through with the balls may pass. It is fed by hand, though in action it is proposed to feed it with a shovel. It is said that a shot from this gun will cut off a nine-inch scantling at the distance of half a mile. The balls used are round, and weigh over three ounces. We have one of them at this office, which was found on the machine at the time of its capture. The gun does not meet with much favor among military men, unless it can be used in a fort or in some place where it could be worked with safety. A six-pound shot fired into the machine, would completely disable it and render it unfit for use. Our southern neighbors, however, seem to think better of it, as we see by papers of yesterday, that another is to be built at a cost of $5,000 and delivered in Richmond. If it is built in Maryland, perhaps, it will get there.

"The machine is not yet on exhibition," the article continued. "It will soon be formally presented to the Mechanic's Association, and will then be exhibited to the public. There is some talk of a public demonstration on the occasion, and a parade of the four military companies here as a battalion, but as they have been deprived of their muskets that part of the ceremony will have to be dispensed with."[113] Notice that the gun had arrived in Lowell, and was to be presented to the Middlesex Mechanics Association, appeared in the August 16 issues of the *Baltimore Sun* and the *Philadelphia Inquirer*.[114]

September 13, 1861, found the gun in the *Lowell Daily Citizen and News* again. "That Steam Gun— This machine was exhibited upon the fairgrounds yesterday, and received a very close examination from some of the best mechanics, who pronounce it to be a great humbug." By November it was relegated to being displayed at the North Middlesex, Massachusetts, Cattle Show among the normal plowing contests, displays of animals, vegetables, fruits, and agricultural implements.[115]

But it had also entered formal, if prematurely, written history with a mention in Evert Duyckink's *National History of the War of the Rebellion*: "There was some talk of making the engine serviceable in the national defense; but we may presume there were inherent difficulties in the way. . . . The 'Winans' Steam Gun,' after supplying paragraphs to the newspapers till all interest in it was exhausted, was transported by way of the Chesapeake and Fortress Monroe as a trophy of war to Boston." Though this account styled it the Winans Gun, it also clearly connected it to Dickinson by repeating his claims for the gun from the circular which both *Scientific American* and *Harper's Weekly* had published.[116]

⇥ 8 ⇤

A Southern Twist

As the saga of the Dickinson Centrifugal Steam Gun unfolded, the Richmond *Daily Dispatch* revealed a chapter of its history previously unexplored by historians. Discussion of the gun appeared in a letter to the editor published on June 23. "We regret to learn that this important instrument of modern warfare has not been taken into favorable consideration by the Confederate Government," lamented a writer who called himself "Tar Bucket." "Its value and importance for destructiveness is beyond cavil. It is also important economically, as there is no necessity for powder, caps, or the many other implements required for a common piece of ordnance." Drawing verbatim on earlier published statements by Dickinson, he continued:

> As a triumph of inventive genius, in the application and practical demonstration of centrifugal force, (that power which governs and controls the Universe, and regulates and impels the motion of planetary bodies around the sun,) this most efficient engine stands without a parallel, commanding wonder and admiration at the simplicity of its construction and destructiveness of its effects, and is eventually destined to inaugurate a new era in the science of war. Mounted on a four-wheel carriage, it is of itself an army, and throws three hundred balls per minute, and is worked by three men. If built on board a propeller vessel, it will throw one hundred 12-pound red-hot shot per minute.

"Tar Bucket" suggested that an "old experienced sea captain" in Richmond wanted to build one to attack the blockading squadron. "If the said vessel is fitted up with private interests, this will be done so soon as a refusal of the Confederate Government, where the proposition is now before them, is obtained." He concluded, "There has been too much delay by our Government officers in bringing these important missiles of war before the Executive. The system of bringing them before a Board of old men, who never recommended anything new, is a bad policy, as the time lost in getting their reports, salaries expended, &c., a gun can be built and brought to use on our enemies. If Government does not adopt it, there will be a fine opening for our enterprising merchants and citizens. The company will then get up in shares of $100."[117]

Reading between the lines, it is clear that an attempt was underway to interest the Confederate government in the gun. It is also entirely possible that "Tar Bucket" was in fact Dickinson or an associate trying to arouse public support for the matter.

An answer to "Tar Bucket" appeared in the *Dispatch* on July 5. "Permit me to express my full concurrence in opinion with your correspondent, 'Tar Bucket,' as to the merits of the Steam-Gun, (the invention of Dickinson,) and of the great impolicy of its rejection by the Confederate Government," wrote an unnamed person from Charlottesville. "The dreadfully destructive powers of this gun, as well as the simplicity of the contrivance — the almost illimitable force of its motive power, and the ease as well as accuracy with which its mighty power as a projectile can be directed, have all been fully tested by actual experiment in presence of the corporate authorities of Baltimore, where it was constructed, and it was at once and eagerly adopted, and the requisite amount for its equipment voted from the city treasury."

Although the gun was tested in Baltimore, the idea that it had been built there and purchased by the city, and that Winans had been involved in the attempt to send it to Harpers Ferry, which the writer also mentioned, have been brought into question, as has his notion, also part of this letter, that the gun was useless because its secrets were known only to the inventor (they were in fact on public record at the United States Patent Office). However, the writer then went in a direction that cannot be proven or disproven from available sources. "It is known that he [General Butler] made the most liberal and tempting offers to Mr. Dickinson to join their cause and render this gun available to them; but being a true friend to the South, and too honest to be bought by the Yankees at any price, Dickinson persistently refused and came to Richmond, and now offers to cast another and better gun than that [which] fell into Butler's hands, at the Tredegar works, and have it ready for service in six weeks, if the authorities will give him authority to do so." The writer suggested it would be "a great blunder on the part of the powers that be if they decline his proposition, which, if a failure ensue, can only involve the loss of a few hundred dollars, whilst if successful, it will prove a more effectual peacemaker and better champion of Southern rights than all the diplomatists and small politicians in the land. I do not know what a board of scientific officers may say about this gun. It is very certain that sensible, practical men, who are good judges, and have seen it work, are warm and decided in opinion that it is a great success, and that one of these guns is more effective than one thousand men, however armed. The opinions of scientific men have almost always been found against all new discoveries. Witness Galileo, Fulton, &c. Progress."[118]

Perhaps spurred on by such glowing endorsements, Dickinson's efforts to gain funding for his invention took an important step forward on July 23, 1861. "The Chair presented the memorial of Dickinson, inventor of a steam gun," recorded the Journal of the Confederate Congress. The matter was referred to the Committee on Naval Affairs for review. There are several cryptic references to a "newly invented implement of war," in the record of a bill passed on August 5 and signed by Jefferson Davis.[119]

Proof that the gun was the subject of this bill is to be found in the Richmond *Daily Dispatch* of August 7, roughly the same time the original gun reached Boston. After recounting Dickinson's escape from Union troops, taking with him the gun's key parts, the *Dispatch* reported that "This gentleman has been in Richmond for some time, exhibiting to the officers of the War Department his models, machinery, modus operandi, &c, of his steam life-destroyer, and as soon as Congress assembled brought the subject before that body. On Monday, we are informed by Mr. Dickinson, [that] the committee, of which Mr. Conrad is chairman, reported favorably, and Congress ordered the construction of one gun. The cost will be about $5,000."[120]

Dickinson's appearance in Richmond made the papers back home in Cleveland on August 17, 1861. "Mr. Dickenson of this city who exhibited the centrifugal gun to many of our citizens before his departure, had made his appearance at Richmond where he was been feasted and feted and what is better has received an order for a second gun, to be constructed at a cost of $5,000," the *Cleveland Leader* reported.[121]

The next mention of the gun in the records of the Confederate Congress occurred on August 23. As part of a bill to provide for naval operations, an amendment was proposed "To construct [a] centrifugal gun, the invention of Charles S. Dickinson, subject to the conditions of the act passed for that purpose, five thousand dollars, which was agreed to."[122] The Confederate appropriation for the gun was also noted in the September 2 issue of the *Baltimore Sun*.

So far, only one additional mention of the Richmond gun has been found. "A queer looking machine, which, we were told, was a 'Centrifugal

Steam Gun,' was carried up Main street yesterday on a wagon, and was an object of much curiosity," the *Daily Dispatch* noted on October 29, 1861. "The colored driver was unable to explain its peculiarities, and we did not endeavor to learn its destination. The machine had a nozzle like unto a blacksmith's bellows." For now, this second steam gun, remains lost in the mists of time, but perhaps further research will reveal more details of its story.[123]

⇥ 9 ⇤

The Gun and Winans
Leave the News

B<small>Y THE FALL OF</small> 1861, Ross Winans and the first steam gun associated with Baltimore were fading from memory. Winans, along with other Maryland officials associated with states' rights or secession, were arrested on September 12, 1861, but released eleven days later.[124] Rumors of plots to take the cigar ship over to the Confederacy were taken seriously enough to station Company F of the 2nd Maryland Infantry, U.S., to guard her during September and October.[125]

With rumors regarding the ship circulating, the *New York Herald* on October 15 likened its shape to that of a Confederate "infernal machine," a submarine built at the Tredegar Iron Works in Richmond and said to be on its way to attack the flagship of the Union fleet off Hampton Roads.[126]

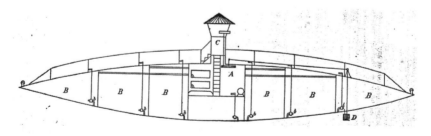

The Winans Tank briefly raised alarm for the U.S. Navy but was allowed to continue on its way once its true purpose was revealed—collecting seawater for engine experiments Ross and Thomas Winans were conducting for future iterations of their Cigar Ship designs. (Image from the Official Records of the Union and Confederate Navies, *Series 1, Vol. 6, p. 348.)*

Imagine the excitement when the tug *Ajax* sailed into Hampton Roads, near sunset on October 19, towing the *Winans Tank*.

The *Tank* was an iron-hulled vessel with a cigar-shaped hull that narrowed to a point at each end. An examination of the ship by Chief Engineer Charles Loring revealed that its hull was divided into six compartments to hold water. A cabin at the center of the vessel provided living space for two or three men. Conversation with those towing the vessel revealed that its purpose was to gather seawater for experiments to determine if the high-pressure steam caused greater build-up of scale—mineral deposits—in a boiler than low-pressure steam. This would be an important consideration for the large cigar-shaped ocean steamer the Winanses hoped to develop, as it would have no masts for use in the event the engine failed.

Flag Officer L. M. Goldsborough, commanding the fleet, quickly reported to Secretary of the Navy Gideon Welles: "The ostensible object is to have it filled with salt water, off Cape Henry, in order to carry out some steam experiments. The construction of the thing is so remarkable, and it could be so easily converted into an instrument of destruction if possessed by the enemy, even by accident, that I thought it proper to forbid its being carried away from here, except by your authority." Goldsborough's report reached Welles after General John A. Dix communicated the *Tank's* business. The *Tank* and the *Ajax* were released and continued on their way.[127]

After the incident with the *Tank*, Ross Winans faded from view in Baltimore. With work on his Cigar Ship brought to a halt by the war, he traveled to Europe, where his son Thomas was working on the next version of the Cigar Ship. But at least one paper suggested another reason for the trip. "The public will learn with surprise that one of the agents for the Rebel iron-clads fitting out in Europe is one of the wealthiest residents of Baltimore," the *Milwaukee Sentinel* reported on November 4, 1863. "He was at one time suspected of sympathizing with the rebellion; but he succeeded in satisfying our government officials of his innocence and soon after started on a 'pleasure trip' to Europe. Of course Ross Winans is the party alluded to."[128]

At the time Thomas Winans was in fact laying plans for another cigar ship, and given this work it is understandable that suspicion would fall on them. In addition to his next cigar ship, the *Ross Winans*, completed in 1866, which was in fact a personal yacht, Thomas unsuccessfully proposed ironclad naval cigar ships to the United States government. His brother William did likewise in Russia but ultimately failed to win over the Russian navy.[129]

⫷ 10 ⫸

A Final Appeal to Mr. Lincoln

By August 1862, the gun had been forgotten in the press of other events. But in light of the accounts linking Dickinson to orders for ammunition for the gun, his involvement in the plot to remove it from Baltimore, and a request for funds put before the Confederate Congress, a letter Joslyn Hutchinson and Charles S. Dickinson wrote to President Lincoln on August 4, 1862, is truly extraordinary. In it they explained the facts of the gun's construction in Boston and its involvement in events in Baltimore, thanks to "the traitor Marshal Kane," who seized it to prevent the passage of "loyal troops." They recounted the attempt to remove it from Baltimore, and claimed that the gun was then found to be in a "damaged condition" and "unfit to be repaired." They noted that exhibitions of the gun at Baltimore proved that it would fire "from four hundred to five hundred two ounce balls per minute," with a "horizontal sweep of 120 degrees," and noted that "the aim can be elevated or depressed at will." The gun was "equally adapted" for use in fortifications and operations in the field and clearly was "one of the most destructive agents of warfare yet invented. It is simple and very strong in construction, and is not liable to get out of repair." They continued with a brief description of the gun's iron shield, which would protect its crew of four as the men shoveled shot into the hopper that fed the gun. When in action, the gun's projectiles "had the same destructive effect when directed against masses or lines, as those projected by gunpowder." The field version of the gun, to be drawn by horses, would weigh two tons when the boiler was filled with water, and for $10,000 Hutchinson and Dickinson offered to build one to be delivered in New York three months from the reception of the order. Their proposal was accompanied by a copy of the May 25, 1861, *Scientific American*, which featured the gun and provided details of its construction but also connected it to Winans.

Interestingly, the letter mentioned another prominent locomotive builder. "As to the value and efficiency of the invention, we respectfully refer to Mr. M. W. Baldwin of Philadelphia."[130] It is unclear if Dickinson was name dropping or if Baldwin was in fact familiar with the operation of the gun, but his tactic does raise the question of whether Dickinson had earlier hoped to legitimize his invention by associating it with Ross Winans, who might have been snared into involvement with the simple question: "What do you think of the gun?"

No response to the proposal has been located, and it is likely that it went no further than the War Department's correspondence file, where it remained before being transferred to the National Archives. Nevertheless, Dickinson apparently continued his attempts to sell his gun. A November 12, 1862, letter from Dickinson to Georgia's governor Joseph Brown surfaced at an auction in 2006. In it Dickinson offered his inventions in defense of the Confederacy. The text of the letter, as transcribed by the auction house, reads:

> Dear Sir I see by your message that you recommend preparing for defence soon as possible, I will send you a few Editorial remarks from men who have Witness[ed] the operation of a piece of Machinery that I have invented for War purposes of late. I am now at Work preparing a Steam Wagon to run on common roads, and will have it in operation in two Weeks and with it I can run round any flying artillery every 20 rods[?]. My address will be at this place for the next 2 Weeks after that time I will be Exhibiting this Machainery. I would like to come your State and Build the first Machine for Servis. Most Respectfully Yours Chas. S. Dickinson. N.B. thar is no difficulty in throwing any Medium sise Ball beyond the reach of Powder.

According to the auction listing, Dickinson included three newspaper clippings regarding exhibitions of his centrifugal gun in Columbus, Cleveland, and Dayton, Ohio.[131] No record of additional efforts to market the gun has yet been located, but Dickinson's mention of a steam carriage raises the possibility of yet another interesting invention. Whether Dickinson was able to lure others into funding another gun, and details of his steam carriage, remain undiscovered.

⊰ 11 ⊱

Doomed from the Start

Despite Charles Dickinson's claims for the steam gun, it failed to introduce the new era of warfare he imagined. Even as the gun was being featured in publications across the United States, the gentlemen of the American Institute's Polytechnic Association gathered in New York on May 9, 1861, to discuss scientific matters, including centrifugal guns. The discussion focused on a different gun, probably that of Robert McCarty then being demonstrated in New York. Several points touched on the problems that doomed the Dickinson gun and others.

Inadequate Range

A Mr. [Frank] Dibben suggested that "the great difficulty in the use of the centrifugal guns is to give the balls the desired direction without impairing their momentum. The balls will all have the same direction, if the barrel is moved, horizontally; that is, they will all strike in a horizontal line; but the lateral deviation will be very great and uncertain, depending upon the velocity of the ball at the moment of discharge. It, therefore, becomes incapable of efficient service." He had witnessed a test of a disk-based centrifugal gun in which the first of a series of shots "went through an inch plank, at the distance of one hundred yards, but the last ones only indented it." Dibben also noted that "Every precaution is now taken in fire-arms to secure accuracy of aim. It is not now a question of one hundred yards, but of three or four hundred yards."

Vulnerability to Enemy Fire

"The steam gun is objectionable," said Dibben, "not only on account of its inaccuracy of aim, but for its want of portability, and for its vulnerability. The boiler is a large mark, and a single shot from a fair distance would entirely destroy its operation." A Mr. Nash, probably Professor. J. A. Nash, concurred, saying that a "steam gun could not be effective unless sheltered, for the moment a cannon ball touches it, there is an end to it. He described the execution which could be done with cannon, or even with rifles, by having a telescopic sight."

Insufficient Force

Dibben noted that disk-based centrifugal guns imparted a rotating motion to shot, which reduced velocity. "Military authorities say that the bullet should have an initial velocity of 12 to 1,500 feet. With light rifles the speed probably reaches 2,000 feet. With such speed as that, the centrifugal gun falls into the background."[132]

Strength of Gunpowder versus Steam

The discussion then turned to gunpowder. A Mr. Stetson noted that improvements in weapons allowed for throwing shot farther with less initial force. Thirteen to fourteen ounces of powder would send a 12 lb. Hotchkiss shell from half a mile to a mile. "The expansive force of gunpowder is very great, having been shown sometimes to amount to 100,000 lbs. [an editor's note corrected this figure to 500,000] to the square inch," Stetson said. "No such force as that of gunpowder can be produced by steam. Nor can we produce the centrifugal force due to such pressure without destroying the machine." Stetson then explained why velocity was a key limitation of centrifugal guns:

> In order to throw a projectile with a given velocity, a force is required equivalent to lifting it a certain height, independent of friction, for that velocity will be expended in raising it that height; or, inversely, if it is allowed to fall from that height, in a vacuum, that velocity will be produced. To produce a double speed in any mass requires a quadruple power, the power required being in the ratio of the square of the velocity. It is thought by some that we may be able, with centrifugal guns, to throw bullets at the initial velocity of six hundred feet, which will kill at short ranges. But we cannot throw a stream of bullets like water through a hose pipe, with that velocity; for merely to overcome the in-

ertia of the balls, a horse-power will only throw 1 1/4-ounce balls at the rate of one per second; and we must have several horse-powers to overcome the friction. If the machine is small and convenient, the number of bullets thrown must be necessarily very small.[133]

Fuel & Water Use

Though not mentioned in the discussion at the Polytechnic Association, using a steam-powered weapon in combat would require a convenient supply of considerable amounts of fuel and water. Though combustible material of all sorts might be easily found, having sufficient water available would always be a challenge.

Artillery pieces of the day were by no means light, but they were portable and packed devastating power. They required but small amounts of powder to throw shells over long distances, and in the hands of trained gunners they were impressively accurate. They had no boilers to leak, pipes to break, gearing to jam, or complicated moving parts. In short, they were brutally efficient at their work.

Though many inventors explored the idea of a centrifugal gun in the days prior to the Civil War and after, their efforts would come to naught,

Conventional field pieces, shown here in Washington during, or just after the Civil War, were moved more easily into action and were deadlier than any steam or centrifugal gun. (Photograph by Andrew J. Russell / Library of Congress.)

as a new generation of gunpowder-based, rapid-fire weapons began to emerge. These guns—the Coffee Mill Gun, and the ultimately more successful Gatling Gun, marked the beginning of a new and more deadly era in warfare, the birth of the modern machine gun.

But though the new technologies were revolutionizing warfare, the idea of a "centrifugal" gun would not die. "This gun was tried in our late war, failed as a moment's thought would have shown that it must, and is now having its day over again in Europe, and attracting considerable notice," *Scientific American* reported on July 6, 1867. In comments that could apply to the Dickinson gun and others, its editors thought that: "It [an unspecified gun design] proposes by the turning of a crank, to hurl sixty to one hundred death dealing bullets a minute. The question is, where is the power to come from? It seems to be supposed to reside somewhere in the crank, the gearing, or the balance wheel. It is the old inexpungeable dream of creating power out of leverage."[134]

Undeterred by criticism, inventors down to this day continue to bring forth proposals for "Centrifugal Guns." As of yet, they have yet to revolutionize warfare.

❧12❧

The Gun Fades into History

ALTHOUGH KNOWN AT THE TIME, the gun's journey into the possession of the Middlesex Mechanics Association in Massachusetts soon faded from memory, and from history. As the years passed and the horrors of war faded, papers began reprinting images from period publications and dispatches. Fifty years after its appearance in Baltimore, the question of what became of the gun arose once more.

On March 30, 1911, the *Boston Journal* printed an 1861 engraving of the gun from *Harper's Weekly* and sought to learn its fate.[135] The next day, under the headline "Ross Winans Gun Captured by Sixth," the *Journal* recounted information volunteered from callers. One informant confused the late Ross Winans with his grandson of the same name, and had him paying French aviator Hubert Latham to fly his plane over Winans' home in Baltimore in November 1910. A veteran of the 6th Massachusetts suggested that Colonel E. F. Jones deserved credit for capturing the gun. Several wartime dispatches about the gun and the senior Ross Winans were mentioned.[136]

Although the information brought forth by these news accounts was not helpful in tracing its fate, coverage during the next month preserved details of the gun's history that might otherwise have been lost. April 1911 was the fiftieth anniversary of the beginning of the Civil War, the Baltimore Riots, and the gun. As part of that remembrance, a spate of similar articles appeared in various papers across the country, offering recollections of the gun by one William Weaver, who claimed to be the very "Weaver" whose 1861 photograph was used by *Harper's Weekly* as the basis of the most frequently reproduced image of it. One was published on April 11, 1911, in the *Boston*

Journal. The *Baltimore News* of April 12, 1911, printed a similar article, as did the April 19, 1911, *St. Louis Times* and no doubt other papers.

Weaver's account seems convincing at first glance but presents many problems. The article painted a dramatic portrait of a then eighty-one-year-old "war artist for *Harper's Weekly*" breaking a "50 year silence" and telling the story of the gun and why it had been found to be unworkable. Weaver claimed that the gun had been built by Winans to "blow its way through Federal forces." Styling one "Dickerson" as the man who had superintended construction of the gun for Winans, Weaver said he removed the trigger, leaving it useless even in the hands of Butler's best gunners. As we have seen, these ideas are completely contradicted by sources closer to the events.[137]

Weaver claimed to have seen Winans "after his gun had been captured, and pronounced a failure by the Federal experts." He also said he had gone to Fort McHenry and taken Ross Winans' picture, had a pleasant chat with him there, and found him to be "an intense Southern sympathizer" who was "very sad over the failure of his plan to deliver his gun in the hands of the Confederates." Given the short time Winans was in Federal custody, the state of affairs at the time, and the day's photographic technology, the idea of a photographer "climbing up a porch pillar" at Fort McHenry to snap Winans' photo is completely unreasonable. The alleged conversation with Winans is contradicted by period press accounts from the Winans family stating that they only repaired the gun, and by the accounting record showing the charge for that repair.[138]

Weaver claimed to have witnessed a test of the gun at Winans' factory sometime in 1861, "against a brick wall about a foot thick," he said. "Heavy timbers, each a foot thick were piled up. When finally placed ready for the test there was about three feet of wood and one foot of brick ready to receive the discharge of the gun. The gun was some 30 or 40 feet away from

The steam gun met its end on the cutting floor of a scrap dealer in Lowell, Massachusetts, surrounded by Civil War veterans who had tried and failed to purchase it. (Photograph by P. V. Ayotte, ca. 1874 / Library of Congress.)

the target. At a given signal an awful roar began. In less than a minute the gun had been stopped. In that short time the heavy timbers had either been smashed or thrown into the air. Every one of us was convinced that the discharge would have mowed down a whole regiment."

Weaver then presented a detailed, but incorrect account of the workings of the gun, which he said he had personally inspected. He described its key operating part as a rotating set of "cup"-like attachments that threw the shot out at high speed. He also claimed that the gun was steam-propelled. His description of the weapon is not supported by period sources, and the notion of the gun being self-propelled is not supported by period illustrations, which clearly show horses attached to the gun.

Given the number of things Weaver got wrong about the gun, his accounts of the conversation with Winans and the test of the gun are both highly suspect. But if Weaver's account only complicates matters, its publication in the April 11, 1911, *Boston Journal* brought forth much more useful information.

The next day, in the *Boston Journal,* under the heading "Sold as Junk Famous Gun Found Way to Scrap Heap," much of the gun's history came to light, thanks to veterans of the 6th Massachusetts and citizens of Lowell. The article included the account of the test at Relay from James Whitaker, mentioned earlier, and word from W. H. H. Mallory, who upon examining the gun at Relay in 1861 recognized speed gearing he had made in Boston the year before. Most importantly, the article answered the question of the gun's postwar fate, which though discussed in contemporary papers, had since been forgotten.

"If I remember correctly," Clark Langley recalled, "it was given to the Middlesex Mechanics Association and exhibited various times until it

became a white elephant." Hamblin Gardner added "that it was often referred to as a toy gun." "The balls used . . . were about the size of marbles
and were fed in to the barrel through a hopper, but there was not sufficient
force in the machine to propel the balls very far." Gardner maintained
that "a person standing not very far from the supposed death dealing
machine would not have had very much trouble in catching the bullets in
his hand." The gun was studied by Civil War veterans and others "interested in military science" when first brought to Lowell. It attracted attention after being stored at the Merrimac Manufacturing Company but
eventually was sold for scrap to the H. R. Barker Company (perhaps
soon after the Middlesex Mechanics Association disbanded in 1899).[139]

"There was not a Civil War veteran in Lowell but felt its loss keenly . . . at
the time it was destroyed many crowded the . . . old machine shop, and its
parts brought exceptionally large sums of money from old soldiers," the *Journal*
added. "Each one that witnessed the sale . . . felt that one of the country's most
historic relics was being annihilated." Lowell Grand Army of the Republic
men had tried to save it but could not. "Many of its pieces are today, however,
saved by the old veterans, and the history of the gun promises to go down in
the annals of the Civil War."[140] Over the years, the *Boston Journal's* prophecy
has come true: the gun has gone down in history, albeit as the Winans Steam
Gun, and the idea of steam or centrifugal guns was explored long after.[141]

Though long gone, the gun's story continued to be retold in the press. In
February 1935, Reginald Winans Hutton, great-grandson of Ross Winans,
told a Baltimore paper, "Ross Winans did not invent the centrifugal steam
gun. He repaired it for its inventor, Charles S. Dickinson of Cleveland,
Ohio. He also made balls for it, at the behest of the Federal government,
under the orders of Gen. [I]. R. Trimble. The pikes manufactured at the
Winans shops were issued under the orders of Marshall Kane and General
Trimble. They were issued for the use of and supplied to, the Federal government." Hutton unfortunately confused the involvement of Baltimore's
Board of Police Commissioners with that of the federal government and the
allegiance of Trimble, who sided with the Confederacy.[142]

Of the various articles over the years, the most complete was to be found in
the pages of the *Baltimore Sun*. Louis Bolander's "The Mysterious Baltimore
Steam Gun" correctly cited Charles Dickinson as the inventor. After recounting Dickinson's wild claims for the gun, Bolander noted that it had become
known as the Winans Gun "because it was believed to have been built in the
shops of Ross and Thomas Winans." Bolander recounted the gun's placement
at City Hall following the riot, along with an account of its capture. According
to Bolander many people, including government ordnance experts, examined
the gun while it was at Relay before being shipped north. "No trials of the gun

here are recorded, but evidently some were made, and proved unsatisfactory, for General Butler became convinced that as a weapon of war it was worthless and had it shipped to Boston as a sort of relic and no more was heard of it. Certainly it played no further part in the Rebellion."[143]

In 1956 the gun received notice in Robert V. Bruce's *Lincoln and the Tools of War*. In mentioning the gun's capture by Butler, Bruce wrote that "Ross Winans, fiery pro-Southern Maryland millionaire, was popularly supposed to be that gun's inventor and manufacturer. Actually it was made in Boston under the direction of the true inventor, Charles Dickinson." He described the gun as being served by four men, and his description of the gun's shape and workings was reasonably accurate, though he described the barrel assembly as a rotating drum, which was not the case. He also thought that "Dickinson labored under a misconception common to inventors of that time, who seem to have supposed that a battlefield, like a pool table or a bowling alley, presented the objects of attack in a single convenient plane, to be cut down from front ranks to stragglers in one bloody swoop." According to Bruce, Dickinson tried to convince President Lincoln that "the southward journeying of his dizzy device had been the work of secessionist thieves and offered to construct another for $10,000. His offer was not accepted." This account is interesting, because it was based exclusively on period sources, including the files of the Office of the Chief of Ordnance, in the National Archives.[144]

A replica of the Winans Steam Gun built in the 1960s. Author's photograph.

By the time of the Civil War centennial in the 1960s, Maryland histori-an Louis Clark had come across accounts of the gun while researching war events in Howard County for newspaper publication, and with the support of the Howard County Historical Society plans for a reenactment of the gun's capture were put into motion. With the assistance of Mark Handweirk, Clark built a replica from wood and scrap metal. In 1962, reenactors celebrated the "Raid on Ellicott's Mills" with a rifle match between "Confederate" and "Union" teams, an encampment, and a reen-actment of the gun's capture using Clark's replica. Today, that replica still stands alongside the intersection of US 1 and Old Washington Road near Ellicott City, Maryland.[145]

In 1984 the gun received notice in the pages of *Maryland Magazine*. The attempt to remove the gun from the city was said to have taken place under cover of darkness early on the morning of May 11, 1861. The gun, which ac-cording to this account was invented by Dickinson but built by Winans, was said to have been headed to Confederate troops outside Washington, D.C. When Union troops closed in, Dickinson escaped, taking with him a "vital component of the firing mechanism." After it was found to be unusable, the gun was posted by the B&O's viaduct over the Patuxent River. "It disap-peared sometime, during or after the war and was never seen again."[146]

The Relay House and the Thomas Viaduct as they look today. Author's photographs.

A 1989 article correctly had Ross Winans repairing the gun and making balls for it but went on to say that he dismantled it and had it hauled out of town, where it was captured by Union troops who could never test it "because Winans had removed the trigger during transit." Winans was described as having "decided to arm the citizenry" by turning out pikes that he then hid in a private residence. After the pikes were found by Butler's troops, Winans was arrested, then released. Soon after his release he loaded a schooner with weapons, an action that triggered a second arrest.[147]

Again in 1998, the gun received notice, as an "armored, self-propelled steam gun capable of firing a hundred balls per minute, a precursor of the modern tank. It was loaded on a B&O freight car bound for Harper's Ferry but was captured by Federal troops and dismantled." Winans was said to have been arrested for his part in the gun, and imprisoned into late 1862.[148]

The steam gun received a renewed burst of attention in December 2007, when it was featured in an episode of the popular Discovery Channel program *Mythbusters*. The production staff found the steam gun on the webpage of the author of this book while researching a segment on a steam gun drawn by Leonardo da Vinci. After completing that episode, the team returned to the "Winans" gun.

To put the principle of the gun to a test, the Mythbusters team built a device using the same shape as the barrel of the original gun, which was mounted on a steel frame and surrounded with a shield made from a steel barrel. Steam was generated using two hot-water heaters, and funneled through a pneumatic drill attached to the bottom of the barrel assembly. The Mythbusters did not attempt to recreate the complicated timing mechanism of the gun, opting instead for an approach that would release several shot into the barrel assembly at once.

When the device was completed, it was taken to a runway for three specific tests. After misfiring into the shield with one shot, five were loaded and fired, giving the claim of 400 rounds per minute a passing grade.

A second test looked at killing force. At point-blank range the device placed shot into the center of a bust of gel used for ballistics testing, but at a range of only twenty yards the gun produced non-penetrating hits on a pig carcass that would have been painful but not lethal.

In a final test, this one to evaluate Dickinson's claims that the gun had a range of three hundred yards or more, the device was elevated to shoot for distance and successfully threw a piece of shot seven hundred yards before it had to be turned off.

Although the tests proved that the mechanical principles were sound, especially if the gun included a better timing device, the Mythbusters came to the same conclusion as those who witnessed tests of the real gun long ago—it could throw considerable amounts of shot, and would kill or destroy at short range, but its lack of killing force over longer distances made it no match for the simpler but effective powder-based weapons of the day.[149]

Publicity surrounding the airing of the episode, which was titled "Confederate Steam Gun," led to a lengthy article in the *Baltimore Sun* detailing the revised version of the gun's history presented in these pages. Though widely circulated on the Internet, it is unlikely to have much impact in overturning the well-entrenched mythology surrounding the gun. In this way, the myth of the Winans Steam gun is much like the urban legends that clog our e-mail in-boxes. It survives by way of being included in secondary sources that are incorporated into other historical works.[150]

Louis Bolander long ago captured the story's essence. "It came to be known as the Winans gun because it was believed to have been built in the shops of Ross and Thomas Winans. . . . Thomas Winans, however, in a letter published in the Baltimore papers, denied having built the gun, and said that it was only sent to their shop for repairs. But the name clung." No doubt the story will continue to cling, as inaccurate accounts of its history remain spread out through countless publications and now the Internet.[151]

⊰ NOTES ⊱

[1] "Mr. Perkins' Extraordinary Steam Gun of 1824," found online at www
.lateralscience.co.uk/perkgun, and "Electricity on Show: Spectacular Events in
Victorian London," found online at www.fathom.com/course/21701713/session3
.html. The latter site details Perkin's National Gallery of Practical Science.

[2] A discussion of centrifugal versus centripetal found at http://www.newton.dep
.anl.gov/askasci/phy00/phy00305.htm provided the basis for this paragraph.

[3] United States Patent No. 1049, December 31, 1838, Machine for Throwing Balls,
Shot & Etc.

[4] *The Scientific American*, May 25, 1861.

[5] Ibid.

[6] United States Patent No. 1,713, August 3, 1840, Machine by Which the
Centrifugal Force Is Controlled in Throwing Balls and Other Bodies.

[7] United States Patent No. 15,529, August 12, 1856, Improved Machine for
Throwing Projectiles.

[8] United States Patent No. 17,339, May 19, 1857, Centrifugal Battery.

[9] United States Patent No. 21,1209, August 3, 1858, Improvement in Centrifugal
Guns.

[10] United States Patent, No. 24,031, May 17, 1859, Improvement in Centrifugal
Guns.

[11] "Gunpowder Superseded," San Francisco *Evening Bulletin*, June 20, 1861, also
printed in the *Milwaukee Sentinel*, June 27, 1860.

[12] *The Scientific American*, May 25, 1861.

[13] *Cleveland Leader*, May 13, 1861. The following information about the history of
the gun is drawn from a series of charges and counter charges between Joslin
and supporters of Dickinson published in the *Cleveland Leader* between May 13
and 18, 1861. Joslin's profession was given in a letter to editor of the *New York
Herald*, May 20, 1861.

14 *Cleveland Leader*, May 18, 1861. Dickinson's profession was given in a letter to editor of the *New York Herald*, May 20, 1861.

15 *Cleveland Leader*, May 18, 1861.

16 United States Patent, No. 24,031, May 17, 1859, Improvement in Centrifugal Guns.

17 *Cleveland Leader*, May 18, 1861.

18 *Cleveland Leader*, May 15, 1861.

19 United States Patent, No. 24,997, August 9, 1859, Improvement in Centrifugal Guns.

20 "Centrifugal Gun, and Gunpowder Superseded," *Milwaukee Sentinel*, March 5, 1860, additional information from the June 27, 1860, edition. The April 26, 1861 *Daily Sentinel* noted that the brother of Dickinson, of "Albany fame in this city," was the inventor of the gun then drawing notice in Baltimore.

21 Macon, Georgia, *Daily Telegraph*, August 10, 1860, *Weekly San Joaquin Republican*, December 8, 1860, and the *Ohio State Journal*, July 31, 1860.

22 RG 156, entry 994, Stack: 7w2:15/17/05, Box 6, Series: Corr. Rel. to Inventions ("Special File"), File: Class 1C # 276, Letter, dated August 4, 1862, Hutchinson & Dickinson to President Lincoln, and a copy of the May 25,1861, *Scientific American*; National Archives. Hereafter cited as Hutchinson and Dickinson to Lincoln, August 4, 1862, National Archives. One of Dickinson's former associates provided a differing account, saying that the gun had left Boston in the spring of 1861, "and was only stopped at Baltimore to be cleaned and put in order for exhibiting at Washington." See L. J. Rider in the *Cleveland Leader*, May 15, 1861.

23 *Cleveland Leader*, January 3, 1861; *Weekly Wisconsin Patriot*, December 22, 1860, and *Wisconsin Daily Patriot*, December 15,1860; *The Daily True Delta*, December 14, 1860 [Louisiana]. The *Baltimore Sun* also noted the Boston demonstrations of the gun on January 25, with an account identical to that above.

24 Hutchinson and Dickinson to Lincoln, August 4, 1862, National Archives.

25 "The Centrifugal Gun," *Baltimore Sun*, April 22, 1861, intimates that the gun was on display in the city for some time prior to late April 1861. The *Scientific American* would later quote from a circular describing the gun, so it is clear that Dickinson had been trying to promote it with colorful claims.

26 Ibid., April 1, 1861, *Baltimore Sun*, January 29, February 5 and 22, and April 1, 1861. Construction nearby led to the collapse of a wall onto the home of David A. Dickinson at 21 North Street. The kitchen was destroyed, but his family escaped unharmed.

27 The *Baltimore American and Commercial Advertiser*, April 22, 1861, mentions the centrifugal gun of W. D. A. Dickinson.

28 Based on a review of period images of the gun.

29 *New York Tribune*, May 4, 1861.

30 *Harper's Weekly*, May 25, 1861.

31 Mention of the gun has been found in papers from California, Georgia, Ohio, Louisiana, New York, and Wisconsin, to name a few.

[32] "People's Convention on the Position of Maryland," *Baltimore Sun*, February 8, 1861.

[33] "At a meeting of the State Rights and Southern Rights Convention," *Baltimore Sun*, April 20, 1861.

[34] Ibid.

[35] The preceding two paragraphs are based on Charles W. Mitchell, "The Whirlwind Now Gathering": Baltimore's Pratt Street Riot and the End of Maryland Secession," *Maryland Historical Magazine*, 97 (2002): 203–10.

[36] See MS 916, The Winans Papers, Box 10, Vol. 14, H. Furlong Baldwin Library, MdHS, for records of the firm's orders, 1860–1861.

[37] Hutchinson and Dickinson to Lincoln, August 4, 1862, National Archives.

[38] Although clearly the gun was invented by Charles S. Dickinson, this account, and a request for a proposal to make shot for the gun mentioned below, bear the name of W. D. A. Dickinson or simply W. Dickinson. This indicates that both Dickinson brothers were involved in events in Baltimore.

[39] The men later captured sneaking the gun out of town suggested that it was seized by Baltimore authorities during the riots, then returned after calm returned to the city. This version of events is also supported by the account of L. J. Rider, who noted the gun being taken from its owners by municipal authorities to be put in order at once.

[40] "The Centrifugal Gun," *Baltimore Sun*, April 22, 1861.

[41] "The Catonsville Battery, Etc.," ibid., April 23, 1861.

[42] "Preparing 'Dogs' of War," ibid., April 23, 1861.

[43] Trimble to Winans, April 22, 1861, Winans Papers, MS 916, Box 3, Folder 64, H. Furlong Baldwin Library, MdHS. It is important to note that one cannot prepare a proposal to produce shot for a gun without having knowledge of it to make measurements.

[44] "The War – Important Arrest in Baltimore," *Scientific American*, July 13, 1861, and Board of Police Commissioners, Baltimore, to Winans Co., April and May, 1861, MS 916. Box 1, Folder 16, MdHS. Any repairs to the St. Timothy's battery, or records of examinations of the steam gun on or about April 22, do not appear in the board's account, though the gun is mentioned at a later date.

[45] Charles Howard to Winans, April 21, 1861, on Board of Police Commissioners stationary, directs Winans to continue delivering pikes to Trimble, MS 916, Box 1, Folder 4 MdHS.

[46] Details of munitions-making for the board are from a Winans & Company accounting record entitled: "Board of Police Commissioners, Baltimore, to Winans Co., April and May, 1861," MS 916. Box 1, Folder 16, MdHS. Interestingly, none of the munitions orders are recorded in the Winans Company order book at the Maryland Historical Society, in fact nothing is recorded from April 10 to July 6. The page detailing orders for materials for 1861–63 is also missing. See Winans Company Order Book, MS 916, Box 10, Vol. 14, MdHS.

[47] Election information from an online version of Scharf's *Chronicle of Baltimore*, found at http://www.webroots.org/library/usahist/cobmd017.html. Text is from

p. 611 of the original book, and the date of the session from Scott Sumpter Sheads and Daniel Carroll Toomey, *Baltimore During the Civil War* (Baltimore: Toomey Press, 1997), 27. The Winans Guard is mentioned in the April 26 edition of the *Baltimore News American and Commercial Advertiser* under the heading of "The Military Organizations." It was a component of the "Fifth Infantry of the First Light Division," and counted seventy members. Fortune, "Millions for Defence," Richmond *Daily Dispatch,* April 23, 1861, and Guns, "Liberality of Marylanders," ibid.

[48] *Baltimore American and Commercial Advertiser.* May 11, 1861, and Board of Police Commissioners, Baltimore, to Winans Co., April and May, 1861, MS 916. Box 1, Folder 16, MdHS.

[49] Daniel Carroll Toomey, *A History of Relay Maryland and the Thomas Viaduct* (Baltimore: Toomey Press, 1984), 15.

[50] [Adjutant to I. R. Trimble] to Winans, April 27, 1861, MS 916, Box 3, Folder 61, MdHS; Board of Police Commissioners, Baltimore, to Winans Co., April and May, 1861, MS 916. Box 1, Folder 16, MdHS.

[51] *New York Evening Post,* April 29, 1861, cited in Edward Everett, *Rebellion Record* (New York: S. Putnam, 1861), 60, and *New York Tribune,* April 29, 1861, ibid., 98. It must be noted that Northern papers put the worst possible construction on Winans' actions. When arrested by Federal troops on suspicion of treason, he would be released within forty-eight hours—hardly the treatment accorded someone as allegedly "traitorous" as Winans.

[52] *New York Tribune,* May 4, 1861.

[53] For example the *Baltimore Sun,* April 22 and May 11, 1861; *Baltimore American,* May 11, 1861; *New York Tribune,* May 4, 1861; and *New York Times,* May 11, 12 and 13, 1861. New York papers were not disposed to friendly coverage of Winans, who had angered many by his aggressive defense of railroad patents against alleged "infringers."

[54] May 11, 1861, "Examination and Statement of John A. McGee," General Correspondence folder #4, (April 26th–May 15th 1861, Benjamin F, Butler Collection, Manuscripts Division, Library of Congress.

[55] May 11, 1861, "Examination and Statement of John [Stucker?] Bradford, ibid..

[56] May 11, 1861, "Examination and Statement of John A. McGee, ibid..

[57] "Examination and Statement of John A. McGee" and May 11, 1861, "Examination of Richard Harden," General Correspondence folder #4, (April 26th–May 15th 1861, Benjamin F. Butler Collection, Manuscript Collection, Library of Congress.

[58] The *Boston Journal* printed a dispatch dated May 10 claiming that the gun was concealed in a box of shavings on a train. A dispatch dated May 11 corrected that report: "Ross Winans Gun Captured by Sixth," n.d clipping from the *Boston Journal,* MS 916, Box 4, Folder 101, MdHS.

[59] The *War of the Rebellion: a Compilation of the Official Records of the Union and Confederate Armies,* Series I, Vol. 2, Chapter 9, p. 634. Hereafter cited as *Official Records.* Corporal James Whitaker of the 6th Massachusetts discounted Jones' participation. "Colonel Jones had very little to do with it except to carry out the

orders of Gen. Butler, who through detectives, whom he employed, was in close touch with affairs in Baltimore." Whitaker added that Winans designed the gun, and had built it. He also said Butler knew when it left Winans' shop and ordered Jones to capture it. Jones' account is from the article cited as "Boston Journal."

[60] Details of time and action from "The War Movements," *Baltimore Sun*, May 11, 1861. The names of units mentioned were identified using the Index to the *Official Records*, Series I, Vol. 2, p. 634.

[61] "Troops in Washington," *Charlestown [Massachusetts] Advertiser*, May 18, 1861, p. 1, col. 3 [transcription found online at www.letterscivilwar.com/5-13-61c.html]. It is interesting to note that, according to this account, two hundred of the men in the unit did not have muskets.

[62] For example, the *New York Times*, May 11, 1861, and the *Baltimore American and Commercial Advertiser*, May 11, 1861. Details of Hare's exploit are from the *Cambridge [Mass.] Chronicle*, May 18, 1861, p. 2, col. 2. Colonel Jones's report, found in the *Official Records*, Vol. 2, p. 634, confirms that Hare rode ahead and captured the gun and held it with assistance from locals. A review of the statements of the three men to Butler shows Bradford clearly understood what he was doing.

[63] "Examination and Statement of John A. McGee, May 11, 1861, General Correspondence folder #4 April 26th–May 15th 1861, Butler Collection, Manuscript Division, Library of Congress.

[64] "Examination and Statement of John [Stucker?] Bradford,, May 11, 1861, General Correspondence folder #4 April 26th–May 15th 1861, Butler Collection, Manuscript Division, Library of Congress.

[65] "The War Movements," *Baltimore Sun*, May 11, 1861.

[66] *[Baltimore] Daily Exchange*, May 13, 1861.

[67] "Local Matters," *Baltimore Sun*, May 13, 1861. The *Sun* also gave a slight variation to the names of the captured men, that John McGee and "Messrs. Brown and Bradford" had been captured with it.

[68] An early report in the *Baltimore American & Commercial Advertiser* on May 11, 1861, counted Dickinson among those captured with the gun. The *Baltimore American* and the *Baltimore Sun* of the same date have him captured with the gun, but two days later the *Sun* of May 13 reported that he had not been captured as was originally thought but was in Annapolis. It is possible that the papers were fed incorrect information in the hopes of diverting attention from the gun to allow for testing to discover its capabilities. Civil War generals sometimes used the press and their own reports, which were often copied and sent to Richmond by pro-Southern War Department clerks, to pass false information to the enemy. See John W. Lamb, "Pope's Escape from Lee at Clark's Mountain," *America's Civil War*, July 1998, for one such story that was not uncovered until the late 1990s.

[69] *Baltimore American and Commercial Advertiser*, May 14, 1861.

[70] "Explanatory." Richmond *Daily Dispatch*, May 13, 1861.

[71] *Philadelphia Inquirer*, May 13, 1892, reprinting a May 13, 1861, report.

[72] *Baltimore American and Commercial Advertiser*, May 15, 1861.

[73] *Cambridge Chronicle*, June 1, 1861.

[74] *The Bay State*, May 23, 1861.

[75] The 8th New York was at Camp Morgan, six miles from Baltimore, May 15, 1861. Found online at http://www.dmna.state.ny.us/historic/reghist/civil/infantry/8thInf/8thInfCWN.htm.

[76] "The *Baltimore American and Commercial Advertiser*, May 15, 1861, Telegraphic News," in the Richmond *Daily Dispatch*, May 16, 1861.

[77] This would stand to reason, since from their viewpoint his only "crime" had been to make weapons for use by the legal authorities of the city, as pro-Southern as they might be at the time.

[78] *New York Times*, May 16, 1861.

[79] Richmond *Daily Dispatch*, May 17, 1861.

[80] *Baltimore American and Commercial Advertiser*, as quoted in *Scientific American*, June 1, 1861.

[81] Benjamin F. Bulter, *Butler's Book* (Boston: Thayer, 1892), 234.

[82] Dick Nolan, *Benjamin Franklin Butler: The Damndest Yankee* (Novato, Calif.: Presidio Press, 1991), 93–94. Though Nolan sheds light on how Winans' release was secured, he unfortunately includes several bits of steam gun mythology, particularly the notion that it was self-propelled, and invented by Winans.

[83] See Butler, *Butler's Book*, 227–29 for the full text of his comments.

[84] See *Baltimore Sun*, May 13, 1861, *Cambridge Chronicle*, June 1, 1861, pg. 2, col. 2, "The Steam Gun," and the Richmond *Daily Dispatch*, May 31, 1861, for three of many examples.

[85] Report of Brigadier General Benjamin F. Butler, May 15, 1861, *Official Records*, Series 1, Vol. 2, p. 30.

[86] "Letters from the Army – From a Light Infantryman," *The Salem [Massachusetts] Gazette*, May 24, 1861.

[87] Capt. John Pickering, Co. I – Lawrence, Massachusetts. James Bowen Clark, *Massachusetts in the War 1861–1865* (Springfield, Mass.: W. Bryan and Co., 1889), 158.

[88] "A Description of the Winans Steam Gun," *The Wisconsin Patriot*, May 25, 1861, reprinted from the *Washington Star*.

[89] *Boston Journal*, April 12, 1911.

[90] *Frank Leslie's Illustrated Newspaper*, May 18, 1861.

[91] National Archives Publication M492, Record Group 107, Letters Received by the Secretary of War, Irregular Series, 1861–1866, Roll 1, Cameron to Butler, May 20, 1861.

[92] Butler to Andrews, May 20, 1861, in ibid.

[93] *Boston Transcript*, May 20, 1861.

[94] Diary of Corporal Fred Smith, found online at www.letterscivilwar.com/diaries.html.

[95] Clemence to Butler, March 23, 1861, Correspondence folder #4, April 26th–May 15th 1861, Benjamin F, Butler Collection, Manuscripts Division, Library of Congress and Smith to Clemence, May 23, 1861, ibid..

[96] *Harper's Weekly*, May 25, 1861.

[97] *The Scientific American*, May 25, 1861.

[98] *Harper's Weekly*, May 25, 1861. It is entirely possible that Dickinson solicited Winans' opinion on the gun's capabilities to use as an endorsement.

[99] "The Steam Gun," Richmond *Daily Dispatch*, May 31, 1861.

[100] *The Scientific American*, June 1, 1861.

[101] *Cambridge Chronicle*, June 1, 1861.

[102] Joslin to Butler, June 5, 1861, General Correspondence folder #5, May 16th–June 15th 1861, Benjamin F. Butler Collection, Library of Congress.

[103] Robert Williams to Col. Abel Smith, June 11, 1861, Volume 27, Entry 2327, Letters Sent by the Middle Department and 8th Army Corps, Part I, Record Group 393, Records of U.S. Army Continental Commands, 1820–1920, National Archives.

[104] Holmes, Phillip W., U.S. Army, Co. G, 13th New York Regiment, Diary, 1861, Manuscripts and Archives Division, New York Public Library; Williams to Pinckney, June 24, 1861, Volume 27, Entry 2327, Letters Sent by the Middle Department and 8th Army Corps, Part I, Record Group 393, Records of U.S. Army Continental Commands, 1820–1920, National Archives. Another account from John B Woodard, dated on the sixteenth, noted that he had just returned to quarters from helping load the gun onto the propeller *Sophia*. Elijah R. Kennedy, *John B Woodard: A Biographical Memoir* (New York: De Vinne Press, 1897), 43.

[105] *The Scientific American*, June 15, 1861.

[106] "Steam Battery," *The Illustrated London News*, vol. 38, no. 1094, pp. 575–76, June 22, 1861.

[107] "Military Rule in Baltimore," *Baltimore Sun*, June 28, 1861.

[108] "A Startling Disclosure in Baltimore. Seizures at the Old City Hall. Wagon Loads of Ammunition," *Easton [Maryland] Gazette*. July 6, 1861.

[109] "Correspondence of the Associated Press, Fortress Monroe, June 30, 1 P.M.," Richmond *Daily Dispatch*, July 4, 1861.

[110] *The Scientific American*, July 6, 1861.

[111] *New York Times*, August 5, 1861.

[112] *Lowell Daily Citizen and News*, August 8, 1861.

[113] "City Matters," *Lowell Daily Citizen and News*, August 13, 1861.

[114] *Baltimore Sun*, and *Philadelphia Inquirer*, August 16, 1861.

[115] "North Middlesex Cattle Show," *The New England Farmer*, November 1861.

[116] Evert Duyckinck, *National History of the War for the Union, Civil, Military and Navy, Founded on Official and other Authentic Documents* (New York: Johnson Fry & Co., 1861), 1:185.

[117] Richmond *Daily Dispatch*, June 23, 1861.

[118] Ibid., July 5, 1861.

[119] Journal of the Congress of the Confederate States of America, Vol. 1, for July 23, August 3, August 6, and August 23, 1861, Library of Congress.

[120] Richmond *Daily Dispatch*, August 7, 1861.

[121] *Cleveland Leader*, August 17, 1861.

[122] Journal of the Congress of the Confederate States of America, Vol. 1, for July 23, August 3, August 6, and August 23, 1861, Library of Congress.

[123] Richmond *Daily Dispatch*, October, 29, 1861.

[124] Report of Allen Pinkerton, September 13, 1861, *Official Records*, Series, 1, Vol. 5, p. 196 and Wool to Dix, ibid., Series 2, Vol. 1, p. 689.

[125] Benjamin F. Taylor, "History of the Second Maryland Infantry," Chapter 1. Benjamin F. Taylor Collection, MS 1863, Manuscript Division, H. Furlong Baldwin Library, Maryland Historical Society.

[126] Mark Ragan, *Union and Confederate Submarine Warfare in the Civil War* (Mason City, Iowa: Savas Publishing, 1999), 17.

[127] Correspondence regarding the detention of the Winans Tank, *Official Records of the Union and Confederate Navies in the War of the Rebellion* (Washington: Government Printing Office, 1887), Series I, Vol. 6, pp. 346–50, includes a description and a drawing of the Tank.

[128] *Milwaukee Sentinel*. November 4, 1863.

[129] Information about these ships uncovered by Michael Crissafulli is online at www.verneianera.com, and is based on his review of a copy of the original proposal in the possession of a Winans descendant and a letter from Winans to the United States government.

[130] Hutchinson and Dickinson to Lincoln, August 4, 1862, National Archives.

[131] Auction description found online at http://www.invaluable.com/auction-lot/dickinson,-charles-s.-1-c-fz44gitzjf.

[132] Possibly this source refers to a weapon's effective force over distance, i.e., its range, whereas we would now equate speed to a weapon's muzzle velocity.

[133] Annual Report of the American Institute of the City of New York for the years 1861–62 (Albany, C. Van Benthuys, Printer, 1862), 366–68.

[134] "The Centrifugal Gun," *Scientific American*, July 6, 1867.

[135] "Actual Civil War Pictures," clipping from *Boston Journal*, MS 916, Box 4, Folder 101, MdHS. This clipping was identified by means of clues in the text of the next item.

[136] "Ross Winans Gun Captured by Sixth," Clipping from *Boston Journal*, MS 916, Box 4, Folder 101, MdHS. An online source helped date this clipping to March 3, 1911.

[137] *St. Louis Times*, April 19, 1911.

[138] Ibid.

[139] "Sold as Junk Famous Gun Found Way to Scrap Heap," *Boston Journal*, April 12, 1911, MS 916, Box 4, Folder 101, MdHS. The date of the organization's closing is from a list of research collections in the Lowell Historical Society found online at http://ecommunity.uml.edu/lhs/research.htm.

[140] *Boston Journal*, April 12, 1911.

[141] Notice that a centrifugal gun was being proposed in Europe is found in the July 6, 1867, *Scientific American*. A gun invented by W. E. Hicks of New York is discussed in the January 1889 issue of *Manufacturer and Builder*. In 1916 a gun invented by Frank McMillan Stanton received notice in several papers. See clippings from unidentified papers, in MS 916, Box 4, folder 101, MdHS. E. L. Rice's gun design, powered by airplane engines, consumed considerable government energy during World War I. See the *National Academy of Sciences: The First Hundred Years, 1863–1963* (National Academy of Sciences, 1978), 231. Online at http://www.nap.edu/openbook.php?record_id=579&page=200.

[142] "Crimea," Clipping from unidentified Baltimore paper, February, 8, 1935, MS 916, Box 4, Folder 101, MdHS.

[143] Louis Bolander, "The Mysterious Baltimore Steam Gun," *Baltimore Sun*, date unknown, MS 916, Box 4, Folder 101, MdHS.

[144] Robert V. Bruce, *Lincoln and the Tools of War* (New York: The Bobbs-Merrill Company, Inc., 1956), 139–40. The pages of Bruce's book are filled with mentions of other unique Civil War weapons, including the McCarty gun mentioned elsewhere in this work.

[145] The text from a 1960s Civil War Centennial reenactment commemorating the "raid" is to be found online at http://www.rootsweb.com/~mdhoward/centenial .htm.

[146] Timothy Nauman, "Winans' Steam Gun," *Maryland Magazine*, 79 (1984): 39.

[147] Alexandra Lee Levin, "Inventive, Imaginative, and Incorrigible: The Winans Family and the Building of the First Russian Railroad," *Maryland Historical Magazine*, 84 (1989): 50–56.

[148] Wallace Shugg, "The Cigar Boat: Ross Winans' Maritime Wonder," *Maryland Historical Magazine*, 93 (1998): 429–42.

[149] Information based on the Mythbusters-wiki at http://mythbusters-wiki .discovery.com and viewing of Episode 93, "Confederate Steam Gun."

[150] This refers to the ideas of memes, false ideas that act much like viruses, advanced by Aaron Lynch, author of *Thought Contagion* and other works on the study of how false ideas move through society,

[151] Bolander, "The Mysterious Baltimore Steam Gun." This is the first source found by this writer to posit that the gun had *not* been built by Winans. Bolander suggested that the cause of the link was an incorrect perception on the part of the public, which has been more than confirmed by period accounts.

⊰ INDEX ⊱

References to illustrations appear in italics